Web-Based Virtual Environment
in Water Systems

Web-Based Virtual Environments for Decision Support in Water Systems

DISSERTATION

Submitted in fulfillment of the requirements of the
Board for Doctorates of Delft University of Technology and of the
Academic Board of UNESCO-IHE Institute for Water Education
for the Degree of DOCTOR
to be defended in public
on
05 December at 10:00 hours
in Delft, the Netherlands

by

Xuan, ZHU

born in Nanjing, China

Bachelor of Science, Nanjing University of Technology, Nanjing, China
Master of Science, UNESCO-IHE, Delft, The Netherlands

This dissertation has been approved by the supervisor
Prof.dr.ir. A.E. Mynett

Composition of Doctoral Committee

Chairman	Rector Magnificus TU Delft
Vice-Chairman	Rector UNESCO-IHE
Prof.dr.ir. A.E. Mynett	UNESCO-IHE / Delft University of Technology, supervisor
Prof.dr.ir. N.C. van de Giesen	Delft University of Technology, CiTG
Prof.dr.ir. H. de Vriend	Delft University of Technology, CiTG
Prof.dr.ir. F.W. Jansen	Delft University of Technology, EWI
Prof.dr. M.P. van Dijk	UNESCO-IHE / Erasmus University Rotterdam
Prof.dr. W. Wang	Hohai University, Nanjing, China
Prof.dr.ir. G.S. Stelling	Delft University of Technology, reserve member

CRC Press/Balkema is an imprint of the Taylor & Francis Group, an informa business

Published by:
CRC Press/Balkema
PO Box 11320, 2301 EH Leiden, The Netherlands
e-mail: Pub.NL@taylorandfrancis.com
www.crcpress.com – www.taylorandfrancis.com

Picture in the cover is coming from author's 3D terrain modeling. The remote sensing image is NASA image created by Jesse Allen.

ISBN 978-1-138-02475-5 (Taylor & Francis Group)

Summary

Visualisation plays an important role in understanding physical processes. 'Seeing is believing' is not only a proverb, but also a philosophy for people to understand their surrounding world. Scientists and engineers use models to describe and simulate phenomena of the physical world. However, the majority of those models are dealing with abstract numbers, which often makes the understanding procedure difficult and slow. When these models are to be used in Decision Support Systems, time becomes a crucial factor.

How to integrate different models and information sources and present them in such a way that even non-experts can understand their implications in a relatively short time and take decisions – that is the purpose of this research. On the one hand, the accuracy of the information needs to be guaranteed, since otherwise nice computer images may be misleading people's decision. On the other hand, how to interact with visualisation results is also important since different people have different perceptions when viewing an image. They need guidance in (i) understanding model outcomes and (ii) transforming the results into knowledge and understanding. The first aspect requires a rigorous understanding of the underlying modelling systems in order to generate realistic visualisations; for the second aspect, appropriate user interface design is crucial.

Water management is a complex process since it covers a wide range of sometimes conflicting considerations. Therefore, decisions are difficult to make. In this research, decisions on disaster management are considered. There are roughly two phases in disaster management: (i) risk management and (ii) emergency response. The main differences between the two are the constraints on available time and level of expertise of the decision maker. Therefore, the user requirements of an appropriate user interface for decision support systems in disaster management is analysed in this research. Stakeholders were classified into different groups including end-users and model developers. These groups were chosen since many of the traditional decision support systems were sharing the same user interface as the simulation modelling software. This may lead to problems in complex situations under time-constraints, since the systems are often difficult to interact with and time consuming to be used in a decision support setting.

Time is critical in emergency response where scenarios need to be presented as quickly as possible and fast interaction is needed to comprehend the onsite situation. Many of the actors in risk management are familiar with GIS systems, whereas in emergency response they usually are not. The majority of the information in disaster management systems is of a spatial nature so that spatial information retrieval is important for users to convert into their own knowledge. Three types of knowledge on geographical information are: Declarative Knowledge, Procedural Knowledge and Configurational Knowledge. A 2D map is good at describing Declarative Knowledge so that it is suitable for representing general scenarios. Procedural and Configurational knowledge are required mostly in the emergency response phase where a 3D virtual environment is better for representing on-site conditions with more interaction capabilities than the 2D map. These two visualisation styles in user interface are used in specific case studies to explore the main components in 2D map-based GUIs and 3D virtual environment GUIs.

Since 2D maps are a further abstraction of the real world, more symbolic representations are needed to refer to specific objects. 2D structures can be represented in raster or vector format. Using colours and layers are important for 2D maps to show information in a more comprehensible way. Web-based applications for 2D map-based GUIs and some important tips for their design are discussed in this thesis. Some disadvantages of 2D representations can be overcome by using 3D virtual environments that can more easily display real objects in 3 dimensions.

Terrain and surface features, especially buildings, are important components in virtual environments for disaster management. The B-reps method proved suitable for modelling a 3D terrain from DEM data, assuming buildings to have closed surfaces. Texture mapping can be used to give more information on the 3D model features by mapping the image file onto the surface on different sides. With texture mapping, 3D terrain models can show more detail of the area than using the symbolic representations of the 2D case. After mapping the texture of the building on the modelled surface, it becomes more realistic to interpret and determine, in the case of disaster management, windows to escape from.

This becomes useful in emergency response because of the capability to indicate the precise location of e.g. windows in buildings. The combination of 2D and 3D views in a virtual environment can enhance the spatial information that can be retrieved for all three kinds of spatial knowledge. Level of Detail techniques are crucial for accelerating 3D scene rendering and determining how users can interact with the virtual environment. The appropriate user interface for disaster management should be separated from model development, since different process phases and visualisation methods for spatial information should be chosen for different groups of users. In this research, both 2D map-based GUIs and 3D Virtual Environments are explored to reveal state-of-the-art applications in scientific inchdata visualisation.

Samenvatting

Visualisatie speelt een belangrijke rol bij het leren begrijpen van fysische processen. “Eerst zien, dan geloven” is niet alleen een bekend spreekwoord, maar ook een houding van veel mensen wanneer zij de wereld om hen heen proberen te begrijpen. Wetenschappers en ingenieurs gebruiken vaak modellen om fysische processen te begrijpen en te simuleren. Echter, de meeste van die modellen werken met abstracte getallen die niet altijd gemakkelijk te begrijpen zijn. Wanneer deze modellen worden gebruikt in Beslissingsondersteunende systemen zijn ze vaak te traag.

Hoe kunnen deze verschillende modellen en gegevensbronnen het best worden geïntegreerd en hoe kunnen de resultaten het best worden gepresenteerd zodanig dat zelfs niet-deskundigen de betekenis ervan snel kunnen doorgronden – daarop is het onderzoek in dit proefschrift gericht. Enerzijds dient de nauwkeurigheid van informatie te worden gegarandeerd, aangezien mooie plaatjes soms ook misleidend kunnen zijn. Anderzijds vraagt het omgaan met gevisualiseerde rekenresultaten ook aandacht, aangezien verschillende gebruikers een verschillend beeld hebben bij het zien van een plaatje. Zij dienen geholpen te worden bij (i) het begrijpen van modelresultaten; (ii) de resultaten om te zetten in kennis en begrip. Het eerste aspect vraagt een gedegen kennis van de onderliggende concepten en modelsystemen, teneinde realistische beelden te kunnen verkrijgen; voor het tweede aspect is een juist ontwerp van de gebruikersinteractie van essentieel belang.

Waterbeheer is een complex proces dat zich afspeelt over een breed scala aan aspecten die soms met elkaar in strijd zijn. Vandaar dat het moeilijk is om hier beslissingen te nemen. In dit onderzoek wordt de aandacht gericht op het beheersen van rampen. Daarin bestaan ruwweg twee fasen: (i) risico beheersing en (ii) rampenbestrijding. Het belangrijkste verschil tussen beide fasen ligt in de beschikbare tijd om te reageren en de achtergrondkennis van de beslissing nemer. Daarom is in dit onderzoek aandacht besteed aan het ontwikkelen van criteria voor het juiste ontwerp van de gebruikersinteractie met onderliggende systemen. Daarbij is onderscheid gemaakt tussen eindgebruikers en ontwikkelaars van software modelsystemen. In het verleden werden vaak dezelfde interactiemogelijkheden aan beide groepen geboden. Dat leidt echter tot problemen in ingewikkelde situaties waar snel beslissingen moeten worden genomen en geen tijd beschikbaar is om aandacht te besteden aan details.

Tijd is van groot belang bij het leveren van noodhulp waar oplossingsmogelijkheden zo snel mogelijk moeten worden nagegaan en snelle handeling van levensbelang kan zijn. Bij het uitvoeren van risico analyses wordt veel gebruik gemaakt van GIS systemen, bij rampenbestrijding meestal niet. Veel van de informatie die nodig is bij rampenbestrijding is echter ruimtelijk van aard: om welke locaties gaat het? Waar zijn de beschikbare middelen? Allemaal ruimtelijke informatie die verzameld en doorgegeven moet worden en kennis vragen van verschillende processen.

Vaak worden daarbij drie soorten kennis onderscheiden: declaratieve kennis, procedurele kennis en configuratieve kennis. Een platte kaart is zeer geschikt om declaratieve kennis vast te leggen in algemene scenario's. Procedurele en configuratieve kennis zijn meer van belang bij rampenbestrijding waar driedimensionale representaties van de omgeving een beter beeld geven en betere interactie mogelijkheden bieden. In dit proefschrift worden deze twee stijlen van visualisatie nader onderzocht aan de hand van specifieke toepassingen waarbij het gebruik van enerzijds 2D kaarten en anderzijds 3D virtuele gebruiksomgevingen wordt nagegaan.

Aangezien 2D kaarten een grotere abstractie zijn van de werkelijkheid, vereisen zij meer symbolische manieren om specifieke objecten weer te geven. Kaarten kunnen in rooster of in vector vorm worden vastgelegd. Door het gebruik van kleuren en lagen kan nadere informatie worden toegevoegd. In dit proefschrift wordt aangegeven hoe web-applicaties gebaseerd op kaarten kunnen worden gebruikt. Sommige nadelen van platte kaarten kunnen worden opgeheven door gebruik te maken van driedimensionale weergaven die meer overeenkomen met onze echte warnemingen.

Het herkennen van specifieke structuren in het landschap, met name gebouwen, is van groot belang bij rampenbestrijding. De B-reps methode is daarbij zeer geschikt, zij het dat gebouwen daarin als 'dichte dozen' worden beschouwd. Door het patroon van de buitenkant daarop te projecteren kunnen toch de driedimensionale karakteristieken worden meegenomen. Aan de hand van deze afbeeldingen worden de beelden veel natuurgetrouwer weergegeven wat de herkenbaarheid voor de nooddiensten (in geval van rampenbestrijding) veel groter maakt.

Dit wordt met name van belang om de juiste posities van ramen te bepalen en om vluchtroutes te verkennen. Een combinatie van twee- en driedimensionale virtuele omgevingen kan het ruimtelijk inzicht nog verder vergroten voor elk van de drie kennisgebieden. Door gebruik te maken van Mate van Detail technieken waar niet alles in evenveel detail wordt weergegeven, kan de presentatiesnelheid aanzienlijk worden vergroot. Gebruikersinteractie met computersystemen voor rampenbestrijding is vaak niet hetzelfde als voor modelontwikkeling, aangezien het verschillende processen betreft met verschillende eindgebruikers. In dit proefschrift is zowel aandacht besteed aan kaart-georiënteerde toepassingen als aan driedimensionale virtuele omgevingen, teneinde het gebruik van recente toepassingen van wetenschappelijke visualisatie na te gaan.

Acknowledgements

This thesis could never have been finished without the support of a large group of very pleasant people and I feel very lucky to know them. I would like to thank all the people who contributed either directly or indirectly to this work. There are too many names I have to give thanks to, but I would like to specifically mention some here.

First of all I would like to thank my supervisor Professor Arthur Mynett. Without your support I would not have had the chance to pursue my interest in this research on scientific visualisation. Your enthusiasm stimulated me to first obtain my MSc degree in Hydroinformatics and then continue with my PhD research on web-based visualisation. You always taught me patiently and guided me with your broad wisdom and advice. You stimulated me to take part in various workshops, attend conferences and courses to help me understand the state-of-art in my research area and getting to know the scientific community around the world. Together we explored the cases described in this thesis, both in China and the Netherlands.

Prof. Roland Price asked me the question how to prove that my work is useful and it became the leading principle during my entire research. Dr. Yunqing Xuan assisted me in collecting user requirements for web-based interactive map design and we discussed a lot about how web applications should be designed. Dr. Sisi Zlatanova helped me understand 3D GIS and database systems for disaster management. Adrain Almoradie supported me with the server configurations for implementing usability tests. He never said 'no', 'maybe' or 'later' even when he was busy himself with his own research. I would also like to thank Wiebe de Boer for our collaboration in the Building with Nature project. It is nice to work with an enthusiastic person like you who always responded to my emails quickly, even after 10:00 PM at night.

Numerous staff members at UNESCO-IHE and Deltares helped me with various components of my PhD research. I would like to express my sincere thanks to Ms. Jolanda Boots, Ms. Sylvia van Opdorp-Stijlen, Ms. Tonneke Morgenstond-Geerts, Ms. Martine Roebroeks-Nahon, Ms. Anique Karsen, Mr.

Peter Stroo, Ms. Jitka van Pomeren, Ms. Shirley Dofferhoff, Ms. Frances Kelly, Mr. Eric de Jong.

Special thanks go to Dr. Yiqing Guan, project coordinator of the Joint Training Project between UNESCO-IHE and Hohai University. Without your advice I would not have been able to come to the Netherlands to do this research.

I would also like to express my thanks to all my friends in Delft – I am sure I would not have been able to complete this thesis research without their support: Ms. Hong Li, Mr. Qinghua Ye, Ms.Taoping Wan, Mr. Min Xu, Ms. Yuqian Bai, Mr. Zhuo Xu, Ms. Hui Qi, Mr. Leicheng Guo, Mr. Li Shengyang, Mr. Chunqing Wang, Mr. Zhi Yang, Ms. Duong, Ms. Hoang. Special thanks to our 'Weekend Fan Zhuo association': Mr. Xiuhan Chen, Ms. Wen Sun, Mr. Zheng Xu and Mr. Kun Yan.

I am very grateful for the financial support from Deltares for enabling my PhD research. My grateful thanks go to UNESCO-IHE that I already consider as my second home.

Finally, I would like to express my heartfelt thanks to my family. My grandparents always gave me warm smiles and firm belief when I was thinking about giving up. My parents always supported me quietly but strongly. I would like to give my special thanks to my husband Mr. Hui Chen. Without your company and encouragement, I would not have been here now. All thanks go to my angel Miss Yisi Chen – you are all my reasons for striving.

Xuan Zhu

Delft, the Netherlands,

December 2013

Table of Contents

1 Introduction

1.1 Background

Often in life two worlds are distinguished: the *real* world and the *virtual* world. Obviously, the real world refers to the physical world surrounding us and the virtual world to the way our mind is perceiving this: the individual cognitive perception corresponding to the real world. The virtual world can help us understand the real world in which case it is referred to as 'knowledge'. Knowledge transfer is a process to transmit one's virtual world in a way that can be received by other persons. How to display one's knowledge is a key question in all knowledge-related areas. Written text can give a detailed description of one's thinking, whereas images can play an important role because of the visual-stimulation that can provide a more direct and deeper impression. According to neuro-scientists, more than 50% of the human brain is dedicated to processing visual information. With the advent of advanced processor technologies, *computer based visualization* emerged as a special discipline in the 1980's and rapidly evolved in the decades since. Initially computer graphics required dedicated hardware workstations and specialized software packages. However, many advances were made in graphics hardware and software development that lead to more generic and more affordable resources for visualization. The role of computer gaming and movie industry was pivoting for this development – the movie Jurassic Park 'brought dinosaurs to live' .

What is the role of computer visualization in our daily life and what is the impact of visualization techniques? One of the achievements is that users are now able to see things they were not aware of, and these insights help them to define new questions, new hypotheses, new models and new requirements on data (Wijk 2005). Visualization provides a tool that assists people in generating their own 'virtual world'. And, more importantly, this creation can now be transferred and duplicated to help people interact amongst each other and with the 'real world'.

Duclos defined the visualization system in the context of a Shared Data Space and identified 3 main components: (i) Shared Data Space; (ii) Visualization Operator; (iii) Users. He defined the visualization system as a set of Visualization Operators acting in a Shared Data Space, under the control of a User. The different operators in the system are independent. Moreover, the life time of the Shared Data Space and the operator can be different, which means there is no explicit network of operators and new operators can be added to an application dynamically and independently from previous operators within the particular application (Fig. 1.1).

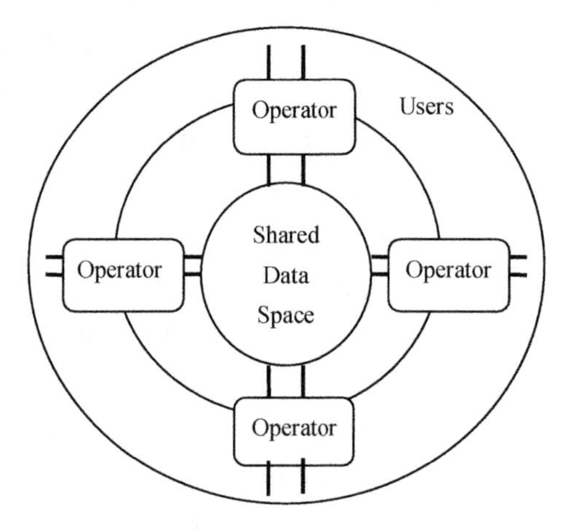

Fig 1.1 Visualisation System Concept

The data objects handled by an operator are called 'Visualization Variables'. Links between operators are realized by sharing some of the data objects of the Shared Data Space (SDS). The behaviour of each operator is controlled by Control Variables. Some of them are made visible to the User and become the (graphical) user-interface with the operators.

The interaction between the SDS and the Visualization Variables are composed of two processes: Import and Export. Import means the Visualization Variables acquire access to the data objects of the SDS; Export means providing access to data objects used by other operators. The interaction between the Control Variables and the Users are a bi-directional process which includes Acquisition and Disposition. Acquisition enables the connection of Control Variables of visualization operators to 'widgets' made available to the end-users for them to perform actions on the application. Disposition facilitates the delivery of results generated by the operators.

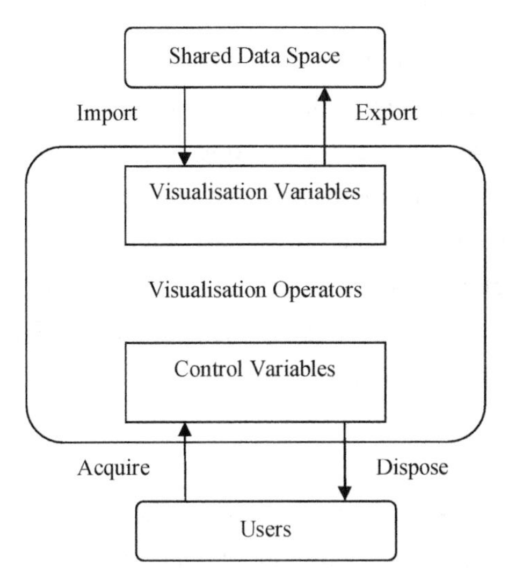

Fig 1.2 Shared Data Space

The visualization operators can also be considered as user interfaces that play an important role in knowledge transfer. Five components are crucial in the whole process: (i) data; (ii) models; (iii) knowledge; (iv) user interface and (v) users. Knowledge plays an important role in the decision-making process since it guides to make appropriate choices. The definition of knowledge here refers to the awareness of the wide range of information related to the topic at hand. This information can be collected from different methods, e.g. from (computational) modelling, from measurements, and so on.

When using the paradigm of a share data space, the 'disposition' process generates the knowledge. The user interface is the visualization tool for representing the information from a data space and also for assisting user interaction with the data space. Therefore, knowledge can be formed in end-users' minds. Users can 'upload' their data into the share data space that then can be 'downloaded' by other users to be shared and interpreted through the user interface to become their knowledge as well. In this way, the important role of visualization in knowledge transfer is clearly revealed.

Process modelling can be considered as an operator which retrieves the requested process information from the shared data space, in which case the user interface is often tailored for a specific model (in the context of this thesis, process modelling mainly refers to 'physics-based modelling' like flood simulation or land-use change). Users of these kinds of user interfaces should already have a relevant knowledge level in order to 'acquire' the control variables. These UI's can be summarized as technical/model interface (TUI) for users with a technical background. One of the main requirements of using a TUI is to understand the underlying physical processes and to be able to communicate with databases and models from different programming platforms and in different programming languages (Lam, Leon et al. 2004).

The complexity of a TUI often corresponds to a particular model schematization. The 'Acquire' mode follows the models' input requirements and assumes that the user can understand the meaning of the various parameter options. The 'Dispose' mode is related to the special processes and to the coupling of different domain models. In case of domain models with different length and time scales, this process is often time consuming and difficult.

In contrast, there is another type of UI named public user interface (PUI) whose users are not specialists and may have widely different backgrounds and knowledge levels that are not the same. This group of users is commonly found as the group of the stakeholders to be addressed by Decision Support Systems (DSS). From studies in several projects on user requirements by stakeholders (Steven 2009; Steven 2010) it is observed that running complicated models is not what they expect from a decision support system since they do not intend to use the particular models in that way. For them, the decision making process is trying to obtain the essential features of the knowledge displayed in a common, easily understandable way. Such gaps between scientist and engineers on the one hand, and managers and decision makers on the other, are revealed in the Frame and Reference Method created for structuring the end-user-specialist interaction in application oriented knowledge domains (Koningsveld 2003). Effective interaction is highlighted to prevent or postpone the seemingly inevitable divergence of end users' and specialists' perceptions on what is relevant knowledge.

A suitable modification of the share data space is to abstract the process modelling from the operators. The visualization part is like an independent mould, which can be loose or coupled to different models. In this way, the visualization with user interaction is separated from the modelling component. Therefore, the focuses can shift to the visualization and interaction with the model results database, rather than mix the model developing and user interface design together. Hence, the problem of appropriate models and appropriate user interfaces is also separated. An appropriate UI is strongly related to the user requirements. In this research, emphasis is on the UI of decision support system in water resources management. The common questions for UI design are the design of Graphical User Interfaces (GUI) and data accessibility.

Two kinds of visualization styles are being discussed a lot: (i) two-dimensional (2D) and (ii) three-dimensional (3D) GUIs (MacEachren 1991; David, Peter et al. 1995; Zhu and Chen 2005; Zlatanova 2008). Map-based GUIs and Virtual Environment GUIs are considered representative visualization tools which are important in communicating spatial and temporal information and interaction (Fig. 1.3).

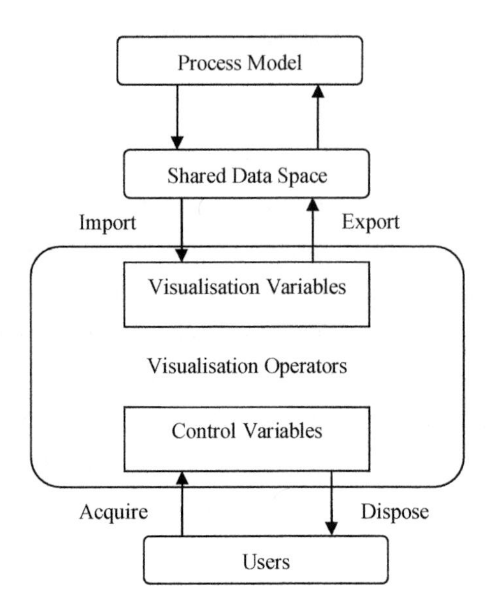

Fig 1.3 Extended Share Data Space with Process Model

1.2 Decision Support Systems in water management

A decision support system (DSS) can be perceived as a combination of human intelligence with computer capabilities in order to solve complicated problems and improve decision making quality. In the 1980s, Sprague (1982) pointed out the classical structure for a DSS, which is based on data and models (Fig. 1.4). From this structure, one can see that the interface is in charge of communication between the decision maker and the information system. Clearly, the DSS interface is neither the model interface, nor the database interface, but rather a separate entity tailored to the needs of the end-user. Water resource management is a complex topic since it includes many objectives in its concept, ranging from flood control to drought relief, from public health safety issues from pollution to adequate water supply for food security. Because of the wide range of issues, stakeholders can include all individuals, groups or organizations that have some interest (stake) in the use or the management of water resources (Leon Hermans 2006).

One of the main characteristics of water management is that the resource is spatially distributed which causes the stakeholders to be spatially distributed as well. This implies that decision-making issues are spatial issues. Hence, spatial planning, resource management and decision making processes involve negotiations and compromises among these numerous stakeholders who typically have different interests, objectives and opinions about how their water resources system should be managed (D.P.Loucks 2005).

In this research, we will not pretend to design any particular decision support system that includes all aspects mentioned above, but we try to extract the user interface as an independent system component and discuss suitable methods and representation styles needed to achieve appropriate interactions in the decision making process. Therefore, we first explore some general requirements for the design of any UI.

From the specific features of water management systems, we extract the general requirements for UI design: Integration, Adaptation and Communication. Integration means multiple data sources in different formats can be entered into the system. Data management techniques and standard protocols for data transformation are highlighted in solving this problem. Using a Common Data Format (CDF) not only improves integrated modelling, but also proves convenient in the information visualization processes. Numerous data standards (e.g. netCDF, KML, XML and so on) play an important role in data dissemination. Adaptation means that users can browse among different scenarios and adjust parameters. Modelling techniques and interaction methods are all important components of adaption mechanisms. Communication means that different stakeholders should be able to make use of the platform to discuss the effect of measures. Internet is the most commonly used resource sharing facility during this decade, so in this thesis we focus on web-based platforms for implementing data sharing and data visualisation.

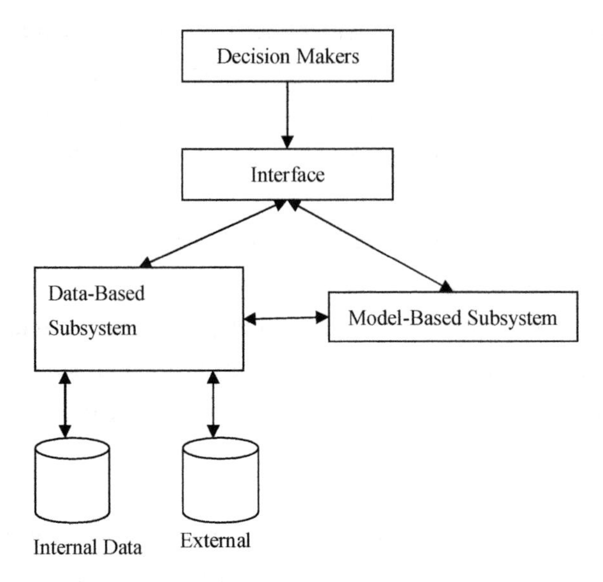

Fig 1.4 Classical Structure of a DSS

1.3 Web-based Virtual Environments

Since the Internet became known to the wider public in the early 1990s, it developed into an invisible network that connects people from all over the world. The value of Internet is not primarily in the technology but also in social communication. Web 1.0 and Web 2.0 are two stages in the development of Internet. Web 1.0 refers to the first stage of the World Wide Web project, which led to the public Internet having as main characteristics the use of hyperlinks to communicate between different web pages. This can be considered as the web-as-information-source which supports the activities of information retrieval. The role of the end-user is in information reading rather than participating. Compared to Web 1.0, Web 2.0 extends the concept from information-centred to web-as-participation-platform. Web pages become more dynamic, showing the ability to "move the power to the desktop" (Robb, 2002). Advanced technologies have been developed for realizing the transformation from Web 1.0 to 2.0, e.g. Asynchronous JavaScript and XML (Ajax). There are three characteristics about Web 2.0: (i) Rich Interact Applications (RIA); (ii) Services-Oriented Architecture (SOA); and (iii) Social Web. Now is a good opportunity to introduce these methods and techniques into different application areas, since the web platform will become dominant in future not only because of its increasing computational capabilities, but also because of creating the environment that support everyone sharing the ideas.

Virtual Environment means the environment created by computer systems and represented through the computer screen. In such environment people can test different scenarios for future planning without side effects to the real world. Indeed, to construct such Virtual Environment requires deep understanding of the real world. This part is handled by modellers and domain experts. This research focuses on the visual display and dissemination part through virtual environments. The representation about virtual environments can be in 2D or 3D or combined. In water-based systems, geographic information is crucial for end-users. Two-dimensional and three dimensional visualisation methods have their own advantages for rendering spatially related information. Using a multi-view approach, the design, decision-making and communication in the system infrastructure design can be supported by an abstract map, a 3D scale model as well as by a very realistic 3D VR scene (Germs, Van Maren et al., 1999). From those virtual environments, end-users can gain more insight about the area and situation that is required to make correct decisions.

Combining web-based techniques and virtual environment technologies provide easy access to study areas. Due to numerous third-party libraries and application programming interfaces (APIs), various functionalities can be achieved. Free and open source software (FOSS) components are main drivers. Implementing these into web-based decision support systems not only cause a reduction in cost, but also increase flexibility of modifying according to customer-defined functionalities.

1.4 Scope of the thesis

The scope of the thesis can be summarized into the following research questions:

- What are the user requirements on decision support systems for water resources management and what are the roles of visualization techniques and styles corresponding to these requirements;
- How can the capacity of web-based user-interaction and display-techniques be used in user interface design for decision support;

- How can virtual environments be constructed for information representation and what approaches are feasible to accelerate the rendering performance;
- What is the role of the 3D information in flood disaster management systems and how can this information be introduced into existing procedures;
- How can user-interaction be improved by advanced visualization techniques and how can these techniques influence decision-making.

1.5 Outline of the thesis

Chapter 2 is on *Appropriate User Interface Design for Decision Support Systems in Water Related Areas – a Focus on Disaster Management.* This chapter analyses the user requirements for disaster management which is one of the important components in a water resources management system. Different users and different objectives lead to different requirements. Visualization styles are identified for 2D and 3D implementations. Basic components, data structures and models requirements are discussed in this chapter. The spatial knowledge retrieval process is described and comparisons made for 2D and 3D visualization styles.

Chapter 3 considers the so-called *One-Page-One-System Design for User Interfaces in Decision Support System.* This chapter explores a web-based 2D map user interface as a decision support tool. As a proxy of a complex modelling system a decision model a Bayesian Belief Network (BBN) is constructed to enable a more rapid evaluation of potential impacts of different measures. The UI design follows internet standards and the concept of Web 2.0 techniques. Clear layout and advanced interaction methods are key components in this design, to reduce the 'jumping windows problem' of conventional web applications.

Chapter 4 introduces a *3D Virtual Environment Information Representation System.* This chapter describes the components and the framework of virtual environment construction. There are two main parts in Virtual Representation (VR): (i) the 3D terrain topography and (ii) objects within/on the terrain. A 3D model of the terrain was built based on Digital Elevation Model (DEM) data. Interaction methods and query functions over the virtual terrain are explored.

The performance of the VR method is a main problem in the system development for large-scale terrains. A so-called Level-of-Detail (LoD) approach is implemented in an inundation simulation process in order to be able to accelerate the decision making process.

Chapter 5 looks into *3D Information in Flood Disaster Management System Design.* 3D GIS is a new discipline that employs 3D information in areas where it is needed most, e.g. for cadastral urban planning, disaster management and so on. In this chapter, a concept model is developed to identify the need for 3D information in flood disaster management systems and how to couple this information into existing management processes.

Chapter 6 focuses on *Visualization and Interaction Technique for Spatial Planning in the Coastal Zone.* A system is designed for representing and analysing results from the Building with Nature (BwN) project. The users of this system include residents and stakeholders living around the study area. The objective of the system is to help public understand the natural process of coastal sediment transport and ecosystem behaviour. Data management and dissemination are the leading questions in this research. A web-based visualization system is developed and user interactions identified for the system design.

Chapter 7 summarizes the *Conclusions and Recommendations* on the capabilities and limitations of visualization techniques in decision support for water resources management. It reiterates the important role of advanced visualization methods to help solving water related problems. Recommendations for future work are suggested.

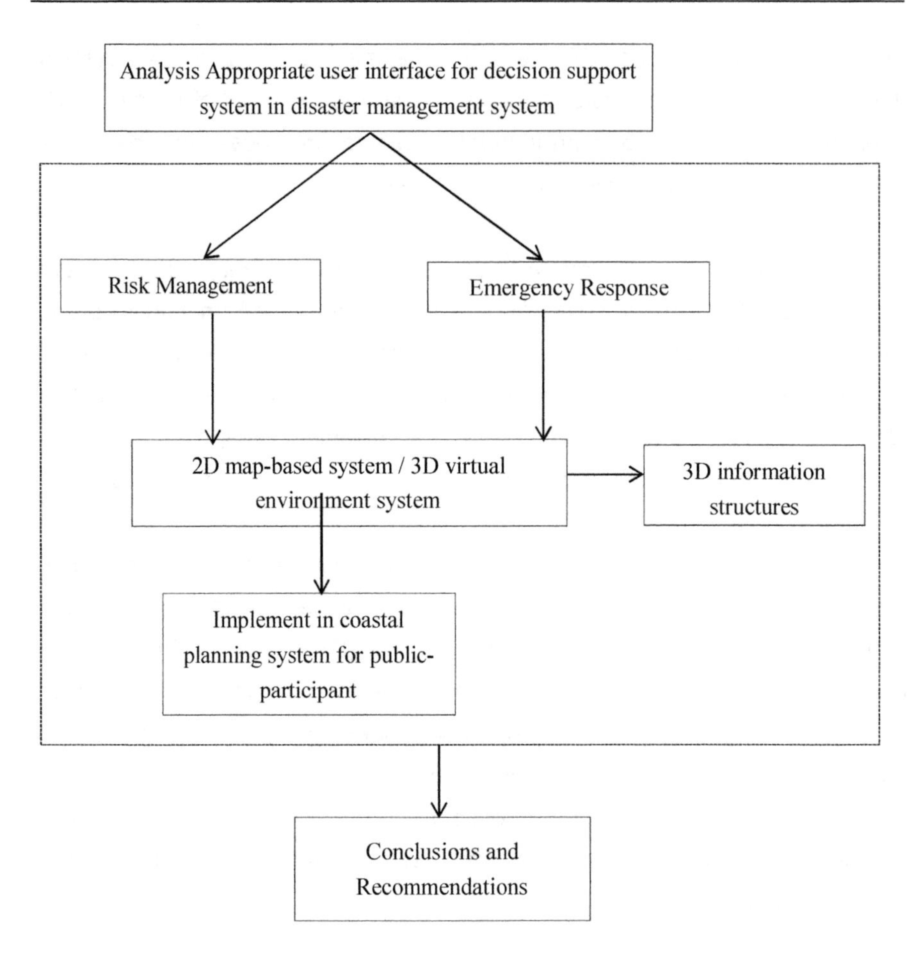

Fig 1.5 Thesis Outline

2 Appropriate User Interface for Decision Support Systems in Water Management

Focus on Disaster management

2.1 Introduction

Efraim (1993) defines a decision support system (DSS) as an interactive, flexible, and adaptable computer-based information system specially developed for achieving solutions for complex management problems. Five elements are usually distinguished in decision support systems: data, models, knowledge, interface and users. The User Interface provides help to users achieving insights from different data sets or model outcomes, which is one of the most important factors to evaluate the system's usability. User requirements are the driving force for user interface design. How to identify the user requirements for a disaster management DSS, is a key question. Two frameworks are introduced here for classifying user requirements: PIECES and Object-orientation (OO). Within the OO framework, the process of disaster management should be specified first and then the key actors should be identified. Performance, Information, Economics, Control, Efficiency and Service (PIECES) is a framework that defines criteria for evaluating whether or not a system has achieved its requirements. From those criteria, performance and information are the main factors taken into consideration here. There are two phases in the disaster management process: risk assessment and emergency response. Time requirements and identification of main actors are two main differences between those two phases. Time is more critical in emergency response where the main actors are non-GIS experts. Therefore, they need more direct methods for information representation. Since the majority of information that needs to be represented is spatially distributed, the spatial information retrieval is the key factor to influence the knowledge creation process. Three types of spatial knowledge (Declarative knowledge, Procedural knowledge and Configurational knowledge) are explored for studying the appropriate representation method for spatial information. Both 2D and 3D visualisation methods are applied to reveal which one is most effective in spatial knowledge retrieval processes. As a result, 2D Map-based systems and 3D Virtual environments have been identified as the preferred user interface design styles for disaster management decision support systems. The basic components and interactive functions are compared in order to find out suitable representation methods correspond to the different phases in the disaster management process.

2.2 User Requirement Analysis

2.2.1 Method of Studying User Requirements

Establishing user requirements is about determining what is "in" and what is "out". The difficulty about the user requirement research is that the requirements are hard to document. Some of the requirements come from non-professionals using natural language to define preferred computer-systems needs. Those requirements are often not easily perceived and need to be interpreted. Most of the user requirement analyses are business-oriented and related to overall analysis of processes in an organization (company, firm, agency) starting from the mission and ending with the final outcome (Zlatanova, 2000). Within the Object-oriented (OO) method followed in this research to study the user requirements, the abstraction of the process can be decomposed into Objects, Responsibilities and Scenarios (Coad and Yourdon, 1990).

First, a clear formulation of the process is needed to identify the objects. Then, the attributes, relationships and services should be revealed between the different objects. Scenarios refer to the sequence of object interactions. In decision support systems, scenarios are important because they determine the direct interaction part in the whole system. Three object model components are distinguished that correspond to the three processes: Problem Domain component, Data management component and Human/System interaction component. Problem Domain component is essential in the whole process since it dominates the other two object models. Although, the object-oriented framework gives a clear procedure to identify the user requirements, it does not pay attention to performance and user interface issues and has difficulties in the evaluation phase. PIECES has the advantage of listing the subjects to be evaluated, e.g. Performance, Economics and Control, which are needed for visualisation and interaction.

Clearly, different groups of people have different kinds of requirements and some parts of the requirements are contradictory. Therefore, we need to classify the users before we analyse the requirements. Norman developed a method to classify the user requirements into three main classes: Global perspective, Individual perspective, Group perspective (Ronald, 1996). Accordingly, there are different methods to collect the requirements for those three groups. From an overall perspective the study method can be decomposed into reviewing available reports, forms and files; researching what has been done by others; surveying similar implementations. From an individual perspective interviews, questionnaires, observations and prototypes can all contribute to determine the requirements. From a group perspective, joint brainstorm sessions, joint application development, joint CASE tool development, etc. all have the advantage that it easily identifies the different interests amongst the users and ways to resolve conflicting issues. The main disadvantage here is how to select the right people for the groups.

Section (2.2.2) mainly explores the various stakeholder groups in the disaster management process and analyse their requirements by using an OO methodology both from a global and individual perspective. Research has shown (Zhu et al, 2005) that the spatial information representation and retrieval phase are the most important issues in user interface design. Section (2.2.3) explains what spatial information retrieval is and how 2D and 3D information contributes to this process. Section (2.2.4) mainly talks about 2D and 3D information representation and their advantages and disadvantages. Two visualisation styles on user interface design for decision support system have been identified in this research and details are given in Section (2.3). Since the concept of 3D GIS was created only one decade or so ago, not everyone is familiar with this application or may be confused by the concept of pure 3D graphic products, e.g. computer games. Section (2.4) identifies the data structures in 2D and 3D models.

2.2.2 User Groups

This research focuses on data and information representation in decision support systems, especially for water disaster management. Therefore, the users of such systems should be identified first. Potentially everyone is a stakeholder in water disaster management. In general, the DSS can be seen as a creation system that produces decisions.

According to (Donoghue, 2002) there are different categories of stakeholders in the system creation process, including End-Users and Software-Engineers. The End-Users category are people who are using the DSS, e.g. the decision makers in disaster management. Their knowledge background can be different and their requirements are typically: 1) Easy to use; 2) Solving the needs; 3) Effective and quick. Software engineers and designers focus on the process simulation and modelling analysis. They depend on different models to generate future scenarios about, in this case, water disaster management issues. For them, the user interface should correspond to the process they want to simulate. A certain complexity of the user interface is needed, however, because software engineers and application modellers often understand the rationale behind the process and they would like to operate the software in a relative short time.

In many case, certainly in the past, the DSS shares the same user interface with the underlying simulation model. However, this often creates confusion between end-users and software developers, due to

- Complex options

Because the simulation systems are based on models, e.g. physics-based models, mathematics-based models, data-driven models etc., many parameters need to be assigned before running a model. Therefore, the ability to set all model parameters should be made available in the user interface to allow adjustment to the model;

- Difficult to interact

Without the knowledge of the modelling system, end-users find it difficult to operate the system and difficult to enter their query questions;

- Difficult to extend

Models have their own fundamental assumptions, which means that a specific model is working on solving a specific problem only. Although, a common model interface has been created for many years already, to deal with a wide range of underlying modelling systems, like OpenMI (Gregersen, J. B., et al 2007), it is still difficult to include everything inside one model.

- Time consuming

Numerical simulation models often cover a wide study area. Due to the structure of the algorithm and the size of the area, the runtime may vary considerably. If a decision support system needs to run a model repeatedly according to the changes in the alternatives, the runtime requirement may become excessive. Although there may be optimization methods to reduce the runtime, the realization of 'one-click, one-scenario' has not been achieved until now. Surrogate modelling has the advantage of having a short execute time, so they can be used to simulate many alternatives in a decision support system. These models are very convenient when developing a decision support system.

Different user categories can be distinguished in the disaster management cycle: Prevention & Mitigation, Preparedness, Response and Recovery. By using time as the criterion, here the main focus is on two phases: Risk Management and Emergency Response. A global perspective is used here for understanding user requirements on risk management and a survey has been done for existing systems on flood disaster management in the Netherlands. The study on user requirements in the emergency response phase was done by investigating individual perspectives from several actors during that phase.

Risk Management

Because decisions on (flood) risk assessment are often related to long-time planning and time is not so critical, many advanced physics-based models can be run to generate the risk maps. Here, most of the actors are usually familiar with geographic information systems and share the same knowledge domain. Therefore, the user interface for such decision support systems can share the same user interface with the underlying simulation system. Although such interface may not be suitable for other stakeholder, it allows having more time for calibration, validation and explanation. The flood disaster management system shown in Fig. 2.1 is such example of a framework for risk assessment and management.

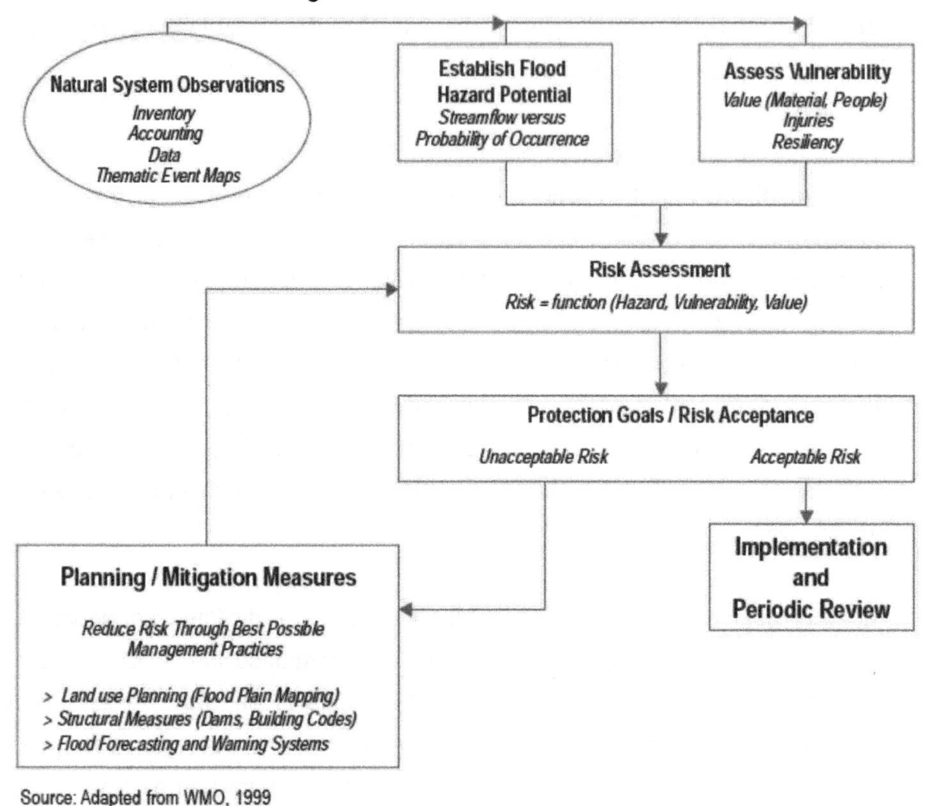

Fig 2.1 Framework of Risk Assessment and Management

From this framework, we can find out the data and simulations that are the main activities in the risk management phase. The data requirements for this phase can be divided into non-real time and real time data. Non-real time data can be considered a relatively stable dataset, e.g. Terrain data, Land-use data, Transportation data, and so on. There are two components in the real-time data class: meteorological and hydrological conditions. They can be detected and filtered by the detection system and be prepared to support the flood forecasting system. Actually, the data correspond to the requirements from the flood forecasting models. Most of the data are represented into 2D maps, if they need to be exported. Most of the users of this kind of data format are GIS users so that 2D data types may satisfy them very well. The main actors and activities in the Netherlands Flood Risk Management Centre are shown in the following table: (Lecture notes TUDelft, 2009).

Activities	Key actors involved
Identification of flood risks	Ministry of Transport, Public Works and Water management, provinces, water boards
Evaluation and assessment of flood risks	Ministry of Transport, Public Works and Water management, Ministry of Housing and the Environment provinces, municipalities, water boards, emergency services, non-governmental stakeholders
Choice and implementation of risk reduction measures and instruments	Ministry of Transport, Public Works and Water management, Ministry of Housing and the Environment provinces, municipalities, water boards, emergency services, non-governmental stakeholders
Monitoring and Maintenance of the acceptable risks	Ministry of Transport. Public Works and Water management, water boards

Table 2.1 The main actors and activities in the Netherlands flood risk management

A survey was carried out on the diversity of geographical information systems within the process of risk management in the Netherlands. The overview of the survey is shown in Table 2.2. In summary, the user requirements for the risk management phase focus on:

- General view of the area,
- Query function,
- Initiating simulation model runs,
- Linking to Rapid Assessment Tools

System Involve	Functions	Main Output
KNMI	Meteorological Database	2D maps, Graphics, Tables
HIS	Geographical Database	2D maps, Graphics, Tables
FEWS	Flood Forecasting	2D maps, Graphics, Tables
Rheinatlas	Flood Risk Mapping	2D maps, Graphics, Tables
ArcGIS	Municipalities Mapping	2D maps, Graphics, Tables

Table 2.2 The general summary of the user requirements of the user interface for the risk management phases

Emergency Response

Time is extremely critical in this phase for the purpose of saving human lives and minimizing property damage. Decision-making is based on information availability on site. Meanwhile, the majority of users, e.g. fire brigade, police, are not geographic information experts. Therefore, the user interface for emergency response support should 1) include onsite information as much as possible; 2) be easy to understand and interact with; 3) not require long runtimes for running any model. Again taking the flood disaster management system in the Netherlands as an example, the main actors and activities are summarised in Table 2.3: (Lecture notes TUDelft, 2009)

Acitvities	Key operational actors
Containment and control of the flood and its effects	Regional fire department. Rijkswaterstaat. Royal Dutch Water Life Saving Association (KNBRD). Royal Netherlands Sea Rescue Institution (KNRM). Military National Reserve
Medical assistance	(Para)medical services (GHOR), Red Cross (SIGMA tesams)
Public order and traffic management	Police department
Taking care of the population	Municipality

Table 2.3 The main actors and activities in the Netherlands for emergency response

(Snoeren, 2007) carried out a survey among 71 users involved in emergency response (fire brigade: 27, police: 11, and municipality: 33). The research results reveal that the demand for 3D visualisation and 3D model results for emergency responders is much higher than 50%. It should be noted that these results are very much influenced by the non-familiarity of the users with GIS so that they prefer another direct representation method to depict the situation as fast as possible. However, although the 3D visualisation features within Google Earth are considered important by some 55%, the most web services for virtual 3D city modelling only support graphics or geometric models, neglecting the semantic and topological aspects of the buildings and terrains being modelled (Kolbe, 2006). It is not enough to only show fancy 3D scenes to the system users, but also present information on details that can support the queries. In summary, the user requirements for the emergency response phase focus on:

- Fast rendering 3D scene
- 3D building information
- Query function over the clickable objects

After having identified the user groups, the most important issue for the user interface design is to find an appropriate information representation method to respond to the specific groups at certain phases. In water related disaster management most of the information is spatially distributed. Therefore, the user interface should maximise the spatial knowledge retrieval process rather than focus on the modelling phase. In the next section, we will discuss the spatial knowledge retrieval process, how to classify the spatial data types, and which information representation method is appropriate.

2.2.3 Spatial Information Retrieve

Knowledge and Information have interchangeable relationships. Fig. 2.2 shows the process of transferring information to knowledge. Knowledge retrieval thus refers to both finding and processing relevant information to generate individual knowledge (Nonaka, 1994). For users receiving information, the way in which this information is presented becomes crucial – and this is where visualisation technologies play an important role. The human eye and brain can process visual information in a parallel manner and a visual interface can help users perceive patterns that might be invisible if information is presented in numbers and tables. Research showed that visualisation also makes solutions more perceptive, since it reduces the cognitive load of mental reasoning and mental image construction which is necessary for internal knowledge generation (Zhang, 1997).

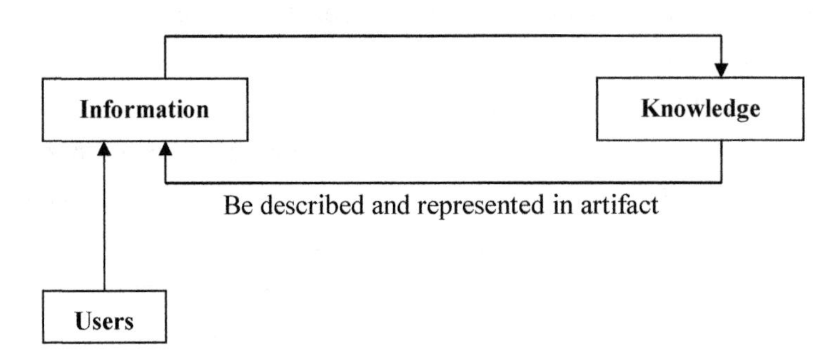

Fig 2.2 Knowledge Retrieve Process

Especially for spatial knowledge, graphical interfaces have their advantage in representing information. Different methods can be used for explaining "where" and "what" problems in multiple images. Take the typical GIS system as an example; the major part of the interface is a big window showing a 2D map. Different information can be represented as separate layers and added onto the surface of the 2D maps. For the technical users, such interface is convenient, because they are familiar with the 2D representation of spatial information. They can easily interpret the meaning of colours. However, general users may not be trained to retrieve this information from 2D maps. People are often excluded from the planning process by difficulties in understanding 2D maps (Richman, 2008). Meanwhile, most of those systems are using multiple windows to display information in different media types which causes a problem when users have to jump from one window to another to obtain the required information (Zhu and Chen, 2005). A 3D interface has the advantage of packing different information in one window. Also, a 3D representation method can easily be understood by many, since it corresponds to everyday life. (MacEachren, 1991) listed three types of spatial knowledge that a user may need to acquire from a geographical information system:

- Declarative knowledge: the information that can be categorized into classes for describing different attributes, e.g. name, location, number of population, etc.
- Procedural knowledge: can also be thought of as routing knowledge, which can be used to show the information between two or more places. A quite common question could be "What is the nearest route to travel from A to B?"
- Configurational knowledge: the cognitive process based on the understanding of the location's relationship and geographical patterns. For example, if we can see that Nanjing is a small point on a global map and that the geographical pattern of this map resembles China, we can conclude that Nanjing is in China. In doing so, we generated the configurational knowledge on Nanjing.

2D maps are good at representing declarative knowledge because of the orthogonal view of the study area with symbols and colours to represent the information. Especially in the risk management phase users are familiar with GIS and 2D map representations. For most of the 2D and 3D simulation models, 2D maps are the front ends of model (e.g. flood propagation) simulations. Generally speaking, such models create the attributes over the 2D maps by providing different information categories, e.g. water level, flow velocity, contaminant density, and so on. Because of the model structure, the information is abstracted into computational units, which are easier to project onto the 2D map user interface for declarative knowledge representation. From research, it was found that the 3D aerial photo was more effective and efficient than a 2D aerial photo in conveying procedural knowledge, whereas 3D aerial photos + 3D semantic maps were more effective and efficient in presenting configurational knowledge (Zhu and Chen, 2005).

A 3D interactive user interface provides more user interaction than a 2D interface, because the users have to rotate the objective many times for achieving the appropriate angles to view a specific area. The procedural and configurational knowledge are more important in the emergency response phase than in the risk management phase, since time is more critical for emergency response where the users are not all geographic experts. A 3D user interface shows its advantage in conveying declarative and configurational knowledge. Procedural knowledge can be shown in a more direct way when the routing information is represented in a 3D style. However, the direct spatial analysis in 3D space is still one of the biggest problems in 3D GIS research, since there is not yet a mature spatial data-model to represent the complex relationships in real 3D space. In this research, two visualisation styles have been explored, corresponding to the two phases of the disaster management process: (i) a 2D Map-based User Interface and (ii) a 3D Virtual Environment User Interface. In the next sections, some details are presented on these two styles. Accordingly, in Chapter 3 and Chapter 4 two cases studies are presented to apply these two styles.

2.3 Two Visualisation Styles in User Interface for Decision Support System

Maps are the most popular representation method for spatial information. The history of the map can be traced to 2500 years ago when people began to draw features on a clay tablet for describing cities. Cartography is the discipline that is all about how to make maps by using abstraction and measurement methods to describe the real world. With the development of information techniques and computer science, paper maps have been digitalized into electronic format that can be disseminated easily. Therefore, Geographical Information Systems (GIS), Remote Sensing techniques (RS) and Global Positioning Systems (GPS) all evolved after the emergence of digital maps. Digital Terrain Models (DTM) are generic models that include both terrain and surface information of the area. The mathematic model to describe a DTM is in essence:

$$Y_p = f_k\left(u_p, v_p\right), k = 1,2,3,\dots,m\,; p = 1,2,3,\dots,n$$

where Y_p represents the NO.p facet on the terrain surface, f_k means the NO.k information, (u_p, v_p) means the coordinates of the NO.p facet, which can be the projection coordinates or the number of columns and rows, m is the number of information types, n is the total number of facets. In this model, both spatial and non-spatial information can be included. For the Digital Elevation Model (DEM) both can be considered as one information type. Both 2D and 3D representation methods are taking this model as their core.

Depending on the different user requirements, we can choose the 2D or 3D representation method in the user interface design, as elaborated in the following two sections.

2.3.1 2D Map-based User Interface

2D maps are widely used all over the world. In many cases, we are reading a 2D map for information extraction, e.g. Google Maps. The reason why 2D digital maps are widely accepted is that it is the extension of 2D paper maps that have already been accepted for a long time. The second reason is that it builds on the mature 2D cartography technique. Also, digital 2D maps already have a standard storage format that can easily be disseminated. 2D maps are an abstraction of the real world that includes symbolic metaphors and projection transfer. It is easier to digitize because a computer screen is also 2D so that projection and scaling can easily be converted. Nowadays, the majority of simulation software is using a 2D Map-based User Interface, e.g. SWAT, Sobek1D2D, MIKE-MOUSE and so on. There are three fundamental components in this style UI:

- 2D data structure
  The 2D data structure is based on 2D computer graphics techniques where mainly two formats are used: vector and raster. For the vector structure, points, lines and polygons are used to represent objects. For the raster structure, the objects are represented by pixels, which are small rectangular cells. The more pixels a graph has, the better the quality of the graph will be. However, the large amount of pixels will increase the size of the storage files.
- Colour model
  Colours provide semantic information to single 2D graphics that extend the information dimensions, e.g. using colours to show the elevation of terrains. With the development of remote sensing techniques, satellite images can support different colour bands for enriching the 2D information so users can differentiate objects on the terrains.
- Layers
  Layers allow users to edit any layer without affecting the others. It is widely used in 2D Map-based User Interfaces for simulation modelling software to assist arranging complex operation on a 2D graph. By

rendering the objects over the virtual canvas separately the users can suppress redundancy when viewing or exporting information.

The main interaction devices for 2D Map-Based User Interfaces are based on the keyboard, mouse and touch screen. Features of the interface are basically about pan, zoom in/out, layers display, clickable items and query functions. There are several basic components of the 2D map user interface:

- Menu: main links to query functions and model assignments
- Tool bar: associated with the main window, always controlling the representation of the map, e.g. zoom in/out, pan, information item selection
- Layer control: the layer structure representing different information categories
- Legend bar: corresponds to the layer control to show the meaning of the layers
- Main window: allocates the 2D maps to the layers selected by the layer controller
- Status bar: shows information about tasks

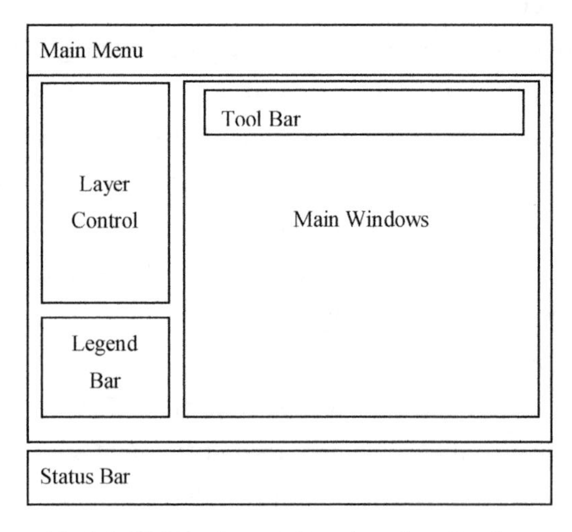

Fig 2.3 2D Map based User Interface Design

Most of the GIS software is using this kind of user interface for spatial analysis and geo-information visualisation. However, the options in the menu can lead the user jumping across various windows to adjust the model settings. Meanwhile, professional terminologies also increase the difficulty of using geographic information software. Nowadays, there is another tendency of using 2D map-based user interfaces. With the development of Internet technologies, many on-line applications for spatial information representation have been developed and some of them affect everyone's daily life, e.g. Google Maps. The advantage of such user interface is that it is very simple and the 2D maps are the main part of the user interface. It is user friendly in that it does not have complex options at the main interface and almost all the information can be selected from the layer controller. The controller is hidden into a small size icon that does not confuse the main window very much. When the users want to view the layer controller, they just need to move the cursor over the icon and then more options can pop up. Another, important feature of this kind of user interface are the clickable items on the 2D maps that users can use to browse for information by clicking on the features on the 2D maps. The main advantage of this setting is that it explicitly links the spatial information with the non-spatial information and the users can have interaction with the information they are interested in. The clickable functions are also very useful when the display device is a touch screen monitor with a map interface. With those features, the interactive activities with the 2D Map-based User Interface can be increased which improves the usability of the software.

The 2D Map-based User Interface can satisfy the user requirements for:

- General description of the whole area for representing Declarative Knowledge by using colour models and symbols
- Eight-direction description for the Routing Knowledge

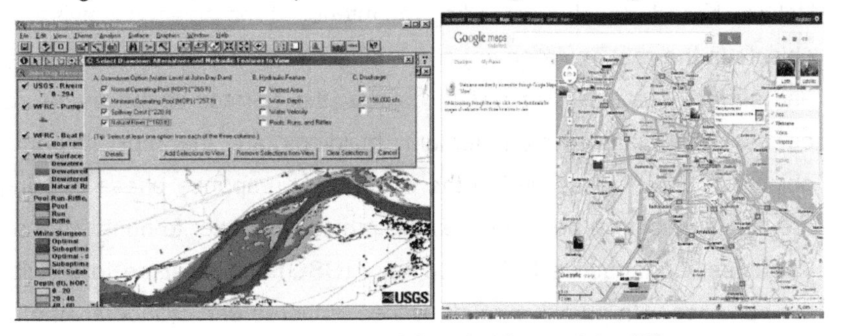

Fig 2.4 Comparison of Google Map and ArcGIS

However, the general problems of 2D visualisation are (Zlatanova, 2006):

- Overlapping Utilities: too many layers on top of the 2D map can cause problems
- Unclear Position: when the positions are located over the elevated area, it is not clear to view the difference in the third dimension.
- Unclear types and symbols: symbols are abstractions of real objects so that it is difficult to standardize unified symbols for every different object in the real world.

2.3.2 3D Virtual Environment User Interface

The reason why to build virtual environments is that it can reduce abstractions of the real world and resume the real objects in the user interface, thereby reducing confusion in human cognition. With the fast development of computer technologies and the specialization trends of 3D GIS, 3D terrain visualisation has become the foreland and focus subject in many areas; it can show the change of a critical area in a dynamic way and can help users navigate in a landscape in order to have an overall view of the area. There are two main components in this style of UI:

- 3D model:

The real objects the model that needs to be represented can be decomposed into simple and complex components in 0D, 1D, 2D, 3D dimensions. The first three kinds can use the 2D structure: points, lines and polygons for representation. In 3D we need to use primitives to model. There are two kinds of primitives for 3D simulation: Surface-based models and Volume-based models. The Surface-based model describes the outside of the 3D object like modelling the 'skin', whereas the Volume-based model is modelling the interior of the object. Likewise, the objects in the real environment can be classified into two classes: terrain and surface features. For terrain objects, the Surface-based model can be used. In disaster management, buildings are the most important objects in urban areas and in the emergency response phase buildings are also very critical because they are providing shelter to human life and valuables. Constructive Solid Geometry (CSG) can be used to simulate the buildings with primitive solids: cubes, spheres and cylinders. Another choice is B-reps that can simulate a building by using a closed surface. The advantage of using B-reps is that it can easier be converted from

GIS and CAD data format, and its primitives are used by most of the rendering engines.

- Texture:

This part is the "cloth" of the 3D model that contains the information of the object rather than its geometrical description. In the terrain simulation part, the texture files is the 2D Maps which includes more information about the area. Different scenarios are represented by different maps, for example the landuse map, the flood inundation maps and so on. It can also be combined with the colour models to show the different attributes upon the area, e.g. if 20% of the highest elevation is water, the users can put the texture on it to represent the water in the 3D terrain. And if the elevation is equal or less than 20% of the highest elevation, the area will be pasted on the water texture which has been chosen. In the surface feature simulation part, the textile files are the photos from different sides of the buildings. After mapping the image on the surface of a specific side of the 3D model, the model will more realistically and precisely show the outside building structure, especially the windows that are important in rescue operation.

The basic components of the Virtual Environment User Interface are almost the same as the 2D ones, but the contents in the legend bar are different because there are no abstract symbols like on the 2D Maps, so that the length of the list can be reduced. Additionally, virtual environments enhance the feeling of the near eyesight scene so that the user interface can present the scene for a general situation, from an orthogonal view of the whole area, to a 3D detailed structure. Therefore, the appropriate user interface for the virtual environment should combine the 2D and 3D view together such that the 2D Map can provide the Declarative knowledge representation and Virtual Environment can support more interaction with variable spatial information. A multi-view approach, combining a 2D plan-view and 3D virtual-view together, provides the best of both worlds by linking GIS and Virtual Reality technology, which is technically possible. (Verbree, E., G. van Maren, H.M.L. Germs, F.W. Jansen, M.J. Kraak, 1999).

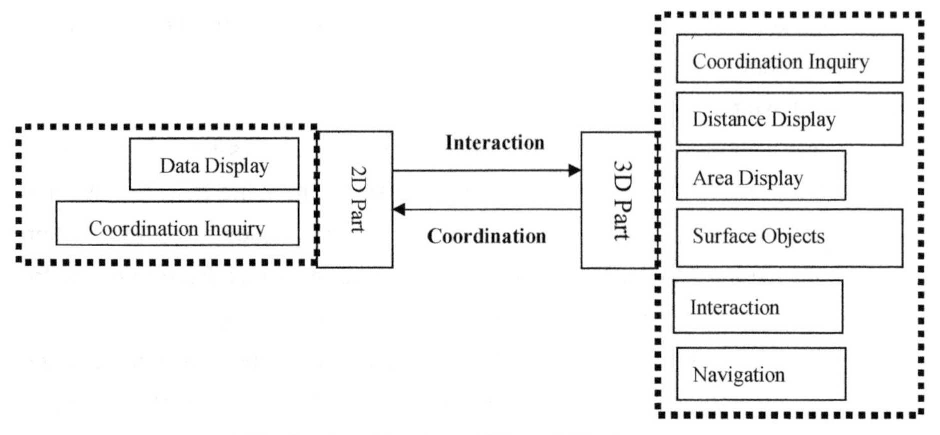

Fig 2.5 Combination of 2D and 3D view

The main interaction devices with the 3D Virtual Environment User Interface are not limited to keyboard and mouse, but also allow immersed navigation using helmet, joy-stick, and so on. However, this research mainly focuses on 'normal devices' by using personal computer as the reference. Options for the interface are basically about pan, zoom in/out, rotation, clickable items and query functions. The basic components are almost the same as for the 2D map-based user interface, but the interaction with the virtual environment is more than in the 2D way because of the additional dimension with the rotational movements. The Level of Detail (LOD) technique is crucial for virtual environments because not only can it accelerate the 3D scene rendering process, but it also determines how users can interact with the virtual environment.

Fig 2.6 Examples of Combination of 2D and 3D view (X.Zhu);(Zlatanova, S)

2.4 Summary

This chapter analysed the user requirements for appropriate user interface design for disaster management decision support systems. Two frameworks for researching user requirements were introduced: the PIECES and OO-approach. PIECES provides a way of classifying problems to extract user requirements and criteria for evaluation. The Object-Oriented OO-method can be used to extract different stakeholders from a description of the process of disaster management. In this research two classes of stakeholders were considered: End-Users and Model Developers / Software Engineers. Early decision support systems were sharing the same user interface with the simulation model developers / software engineers. This lead to problems in case of complex options that created difficulties for end-users, in particular in case of emergency response where time is critical and scenarios need to be presented quickly and more interaction is needed to evaluate the situation onsite.

In risk assessment, most actors are GIS experts since the majority of the information is spatial information. Three types of spatial knowledge related to geographic information are: Declarative Knowledge, Procedural Knowledge and Configurational Knowledge. A 2D map is good at describing Declarative Knowledge so that it suitable for representing general situations in scenarios. Procedural and Configurational Knowledge are required mostly in the emergency response phase where 3D virtual environments are good for representing the onsite situation and for allowing more interaction than 2D maps. Two visualisation styles in user interface were explored and the main components were derived for a 2D map-based GUI and a 3D virtual environment GUI. 2D maps are a considerable abstraction of the real world so that many of symbols are required to represent real objects. Colours and layers are important for 2D maps to display all information. Web-based applications of 2D map-based GUI and important considerations for design were presented. The main disadvantages of the 2D representations are overlapping utilities, unclear positioning and unclear types and symbols.

3D virtual environments require less abstraction of real objects since they can directly show them as 3D images. In this research on flood disaster management, terrain mapping and urban topography description are well represented by surface features, while buildings are important components better described by virtual environments. Texture mapping is able to provide more information on a 3D model by mapping the image file onto the surface at different sides (of e.g. buildings). With the texture mapping, 3D terrain models can show more detail of the area and become more realistic to view. This is very useful in emergency response because it can indicate the precise location of e.g. windows in buildings. The combination of the 2D and 3D views in the virtual environment can enhance the spatial information retrieval for all three types of spatial knowledge. The Level of Detail (LOD) technique is crucial for virtual environments because it can not only accelerate the 3D scene rendering process, but it also determine how users can interact with the virtual environment.

Appropriate user interface design for disaster management should be able to address different process phases and hence contain different visualisation methods for spatial information, depending on the different groups of end-users. In this research, 2D map-based GUIs and 3D virtual environment GUIs are both considered when applying scientific data visualisation to flood disaster management

3 Using Web-based Interactive Maps in DSS

3.1 Introduction

Since the Internet became more widely used in the 1990s, say, it became a driver for many developments, not only in science and engineering, but even more so for social communication. The Internet consists of a huge computer network that connects people all over the world. Compared to communication by telephone, it also provides visual information that facilitates human recognition. Webpages are used for e-commerce, streaming videos, search engines and other new techniques are being created and developed fast, like Social media during the recent decade. Dynamic web paging offers the opportunity to improve the usability of interacting with information, e.g. Google Maps and Google Earth provide spatial information that is easy to access. From the technological point of view, Web 2.0 is a revolution that allows replacing desktop computing with web-based applications. Web 2.0 facilitates participatory information sharing, interoperability, user-centred design and collaboration on the World Wide Web (Paul Graham, 2005). It is not a standard or one single technology, but a group of technologies and frameworks that contribute to a common goal. From a social point of view, using the Internet as the communication and information gathering tool is not only changing the daily life of human beings but also influences the human values. People are nearer to access to knowledge and can communicate more freely than before, although they should be more careful about verifying information in this virtual environment.

Decision support is a complex topic due to the mixture of technologies and social aspects that have to be taken into consideration. Decision-making is a process of making selections and dealing with conflicts between different stakeholders. There are two important components: information and knowledge. Information is based on facts derived from data. Knowledge is about understanding the information. In previous chapters, a detailed description of decision support system components was reviewed and methods were outlined to obtain the user requirements for appropriate user interface design in water (disaster) management. Spatially distributed information is a main characteristic of water management systems. Hence, a 2D map-based user interface for decision support systems provides the starting point in e.g. flood disaster management. Web 2.0 features can support the dynamic interaction between end-users and the decision support system via the Internet. Two parts can be distinguished in the design: the decision support model and the web-based user interface. The work reported in this chapter relates to the EU project AQUAREHAB (VITO, 2010). A DSS was developed for applications focusing on remediation measures in water quality problems. The general framework, state-of-the-art information visualisation methods, and the interaction styles of the REACHER system developed here can be extended to other spatially oriented DSS systems that use 2D maps as their main representation method.

3.2 Background

3.2.1 Project Background

The work reported here was part of Work Package 6 (WP6) of the EU project named AQUAREHAB. The general structure of the project is shown below (Fig.2.1). The objective of WP6 was to develop a collaborative decision support tool, called 'REACHER', that can be used by all stakeholders (citizens, water managers, ...) to evaluate the ecological and economic effects of different remedial actions on water systems.

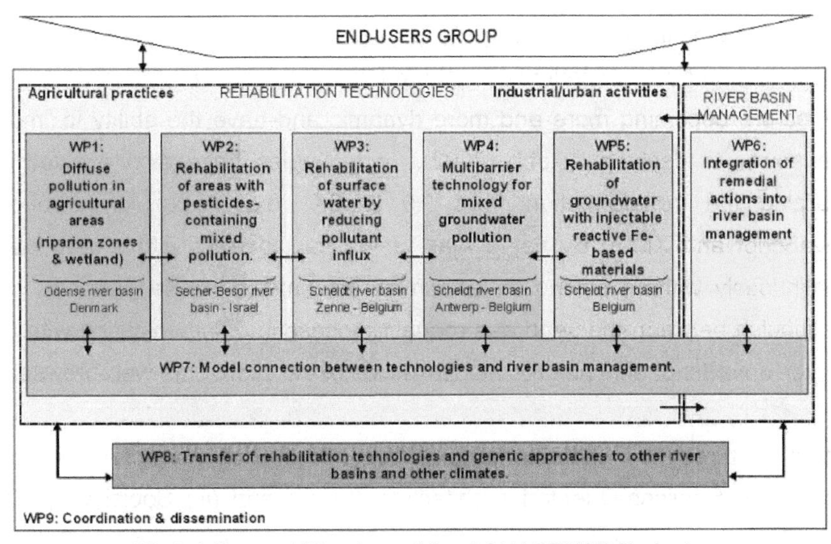

Fig 3.1 General Structure of the AQUAREHAB Project

At the start of developing the decision support tool, there research was carried on identifying the stakeholders and potential users of this tool (Steven, 2009; Steven, 2010). It was concluded that potential users include all levels: national authorities, sub-basin water managers, and public administration. The purpose of the tool was to determine suitable alternatives for future river basin management on

- Impact of measures on water bodies (bio-chemical quality)
- Local vs. central reporting levels: handling at different scales
- Upstream versus downstream impacts
- Possibilities for end-users to interact with the data

Since end-users come from different backgrounds including the general public, complicated numerical model simulations are not the main purpose of the system. The impacts of measures are the main components to take into consideration. Also, a platform was required that can link spatially distributed stakeholders together. Hence, communication via the Internet was an important component in the design.

3.2.2 Internet Opportunities

Two stages can be considered in the development of the Internet: Web 1.0 can be considered as the web-as-information-source stage, which supports the main activities of information storage and retrieval. Web 2.0 extends the concept from information-centred to web-as-participation-platform. User interaction and user experience level are taken into account. Meanwhile, web pages are becoming more and more dynamic and have the ability to "move the power to desktop" (Robb, 2002). Technologies have also evolved for realizing the transformation from 1.0 to 2.0, e.g. Ajax (Asynchronous JavaScript and XML). By using Ajax, users can interact with a web page continuously without having to refresh all the time, because protocols can distinguish between the send and request process in communication with the server. JavaScript enriches the functionalities of the client-side web-browser.

There are three main characteristics of Web 2.0: (i) Rich Interact Applications (RIA); (ii) Services-Oriented Architecture (SOA); and (iii) Social web. The latter has also been referred to as "empowering the entire (world) population as stakeholders" (Abbott, 2001; Abrahart, 2008; See et al., 2008). Web-based applications have the advantage to disseminate knowledge to different groups of users, which is why it has been used here for developing the web-based user interface for the decision support system as a tool for communication and knowledge transfer.

3.2.3 Research Principles

The concept of a 'one-page-one-system' design has long been established for web-based applications. As the name indicates, the system tries to cover all functionalities into a single web page rather than jumping between different pages and windows. In this principle, capturing the users' viewport is the highest hierarchy in the design so that the users can understand their options clearly and not get lost in the system. This idea is coming from the design style of computer games, which emphasises one single scene of the gaming environment, and allocates different options following the player's logic. Especially in Real-Time Gaming (RTG), control is a complex process that includes various decisions under different constraints. However, teenagers (who are often considered having less ability of concentration than adults) can master such kinds of games very well. From the user interface design point of view, this kind of interface holds the users' attention, simplifying options and improving the interaction between users and system. The most obvious feature is the main viewport occupying the major parts of the screen, and elements inside the viewport that are clickable. These settings can guide the users through the system and assist them to retrieve the information they want rather than pushing everything towards them. The one-page-one-system approach is used here to introduce the experience from the gaming industry into the web-based 'water' decision support systems.

The first choice is to select what should be inside the main viewport, and in case of the Vis-Reacher Aquarehab tool, the answer is: a 2D map, following typical map-based systems like Google Maps and Google Earth. In this kind of design, maps, layers, clickable items and options are the main features. At the same time the 'computational engine' selection is also an important factor since this has a direct influence on the database design and computational performance.

Maps are the most important element to support spatial information retrieval and can be used as a platform that can be extended to different information dimensions. Humans have hundreds of years of experience with using maps and this is an advantage when designing a spatially related user interface, like in the Vis-Reacher case. Map-based applications are not a new topic since the concept of Geographic Information Systems (GIS) was established. However, online map-based applications are new topics from the development of Internet-based Browser/Server (BS) architectures enabled by the recent Web 2.0 technologies. Compared to the traditional Client/Server architecture (CS), BS is more convenient because the 'Client Side' are actually web-pages in the browser. References to Web-based Decision Support Systems for water related areas can be found in WaterWare (http://www.ess.co.at/WATERWARE/), Colorado Flood Decision Support System (http://flooddss.state.co.us/Viewer.aspx), SPARROW Decision Support System (http://water.usgs.gov/nawqa/sparrow/dss) and others. There are two kinds of maps used in these DSS: static and dynamic. Maps that play the role of representation tools for displaying different datasets belong to the first group, whereas maps as 'carriers' of interaction options between users and the dataset belong to the second group.

Web DSS applications should be able to represent "what-if" scenarios that can better inform effects of policy decisions than data alone (Graffy, 2008).

Therefore, the combination of static and dynamic map-based systems is important in a Web DSS. Furthermore, the user experience is also an important factor to determine whether or not a Web DSS is a good product. From the survey carried out for the Aquarehab project, there are two main components that contribute to the quality of a webpage: response time and clarity of layout. The first relates to the underlying computational model which is the most time consuming part in the system. The second component relates to the user interface design and data visualisation phase. The three leading questions become:

- How to select an appropriate computational model to save computing time?
- How to combine the static and dynamic map layers online?
- How to design the interaction methods in order to achieve a clear layout?

3.3 Design of REACHER DSS

The REACHER decision support tool was to establish one common decision support platform that can integrate different models, e.g. water quality models, ecology models and economic models. The main topic of the platform is to identify the consequences of remedial measures on specific components and visualize the impacts. Using Free and Open Source Software (FOSS) for developing the environment was also an important component in the development of the DSS. Three types of FOSS tools were chosen, i.e. DSS modeling tool, database tool and web-based visualisation tools. The following table shows the results of the selection.

Categories	Tools
Comuptational Model	SWAT, GeNIe
Database	PostgreSQL, PostGIS
Web-based Visualisation	Apache, Mapserver, Openlayers, Extjs, WebKit

Table 3.1 Selection FOSS software package

For the modelling part, the Soil and Water Assessment Tool (SWAT) has been chosen to construct process-based models for determine the effect of e.g. nitrate and pesticide pollution in the river Odense in Denmark. However, executing the full process-based model takes time and is difficult to modify interactively. Therefore, we chose to implement a lightweight, probabilistic alternative as the computational model to integrate the various complicated process-based fate models. A Bayesian Belief Network (BBN) can approximate the model results in a comprehensive and probabilistic way without high computational demand like in the case of process-based models. It is a suitable management tool for constructing a decision support model. The GeNIe modelling environment and the SMILE graphical user interface (developed by the Decision Systems Laboratory of the University of Pittsburgh) were selected to build the BBN. One advantage of the BBN is that the output of the model can be represented as conditional probability tables (CPT). These tables can easily be transformed into a spatial database so that the user interface can communicate with the model through a spatial database. The graphical capabilities are other characteristics of a BBN since it has the advantage to represent complex topologies. A Web-based user interface was designed to communicate with the database following data transformation standards. Two kinds of layers are developed: static and dynamic. The first one is in charge of communication with the database values and rendering features, the dynamic one contains the computational engine, in this case the Bayesian Belief Network. The interaction style is designed based on a survey of the users' preferences. Clickable items with popup menus are chosen as the principle way of interaction. The process flowchart is shown in (Fig 3.2).

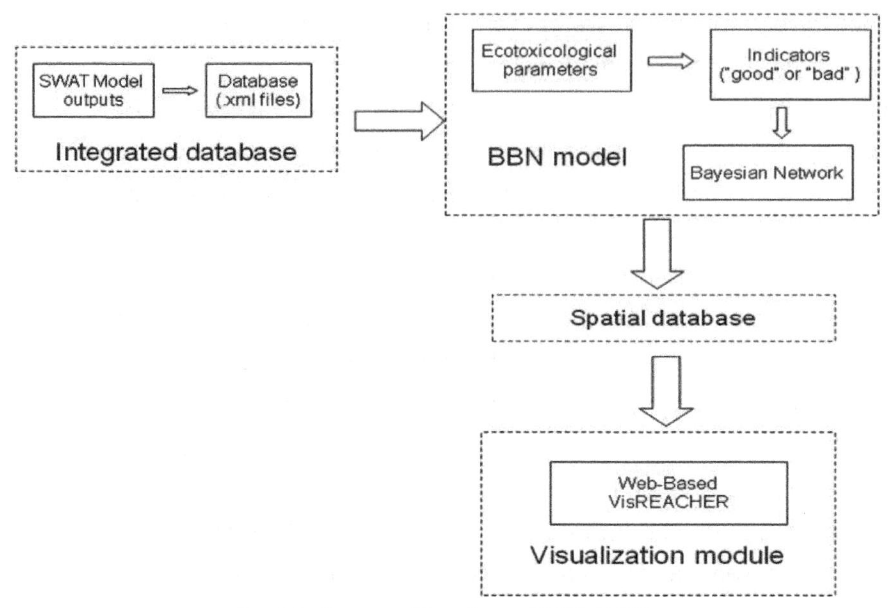

Fig 3.2 Process Flowchart of REACHER

3.3.1 Bayesian Belief Network Design

The fundamental concept of BBN is the Bayesian probability theory, which was established by Thomas Bayes in 1763. The graphical tool designed in this research was to aid the reasoning and decision-making under uncertainty. The BBN network is a directional acyclic graph that uses nodes to represent variables and links to represent the cause dependency. This structure is suitable for the type of water-quality research considered here, because the catchment topology can easily be transformed into a directional acyclic graph, taking the nodes as the sub-basins and the links as the reaches. Within each node, detailed relationships between different elements can be identified. For example, in the case of the Odense basin, the interest is on how nitrogen changes under different measures, and how nitrogen is transported between sub-basins. This process can be represented using nodes and links in a network, following the geographic topology from upstream to downstream. Besides the graphical representation, another important concept is the conditional probability table (CPT). Each node contains such table to establish the relationship with its parents' nodes. The parameters inside the table can be collected from classical statistical analysis or from subjective (Bayesian) statistics. Normally, the frequency of events (e.g. pollution hazards) is not easy to determine because they do not happen repeatedly. However, events can be simulated using process-based simulation models (like SWAT) so that the frequency of the events can still be obtained. A Monte Carlo (MC) simulation approach was implemented in SWAT to generate such frequency of events. Two states were distinguished for events in the statistical calculations: good or bad, because of the requirement to represent the impact of different measures clearly and simply. The rule of distinguishing these two states was based on the ecotoxicological assessment that the 90th percentile of nitrogen was determining the threshold. The procedure to get the classifications from the Monte Carlo simulations is by sorting the values and identifying the 90th percentile of the sorted output. Anything lower than this value is considered 'Good' and above as 'Bad'. The number of 'Good' event was recorded for each component. By following the directional acyclic graph (BBN network), and using the number of 'Good' events in the consecutive nodes which pointed in the direction of the number of 'Good' events in the source node, one can obtain the probabilities of 'Good' in the consecutive nodes when the source node is 'Good'. Then the value can be input into the CPTs. This classification is flexible for modification and it easily implemented in MATLAB (Zheng Xu, 2011). The process flow chart is shown in (Fig 3.3).

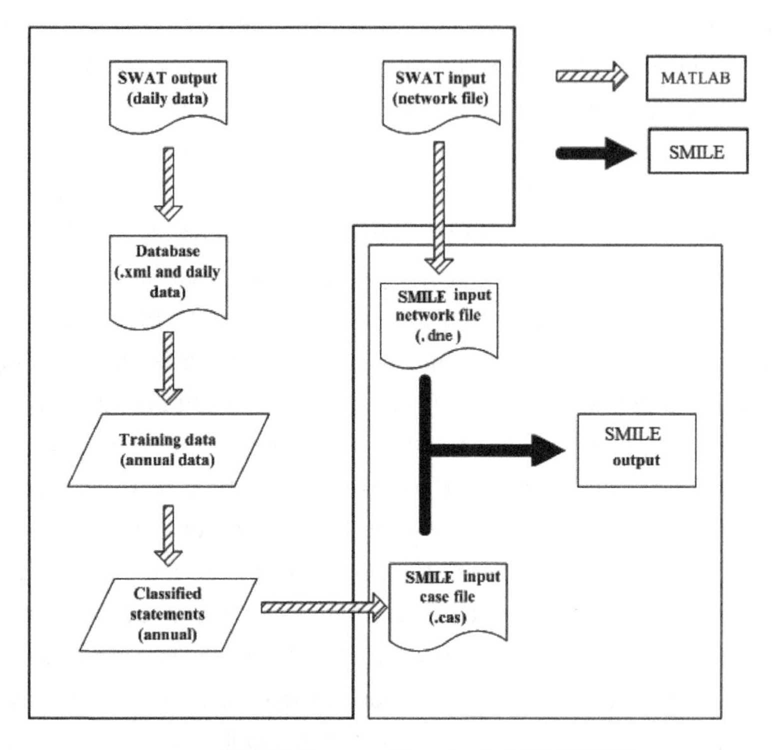

Fig 3.3 Process Flowchart of REACHER

SMILE is a GUI environment for constructing BBN networks and only requires two main input files: the network file and the case file. The network file (.dne) contains the topology of the system as shown in a directional graph. The case file provides the information of the states of the nodes. Once the network is compiled, the CPTs are fixed which means that the initial probabilities for the evidence variables will not change unless the case file is updated. The output of SMILE is also a text-based file named .dne containing the information of the CPTs. The .dne files can be abstracted into table format containing the node identification number and the actual probability value of the node.

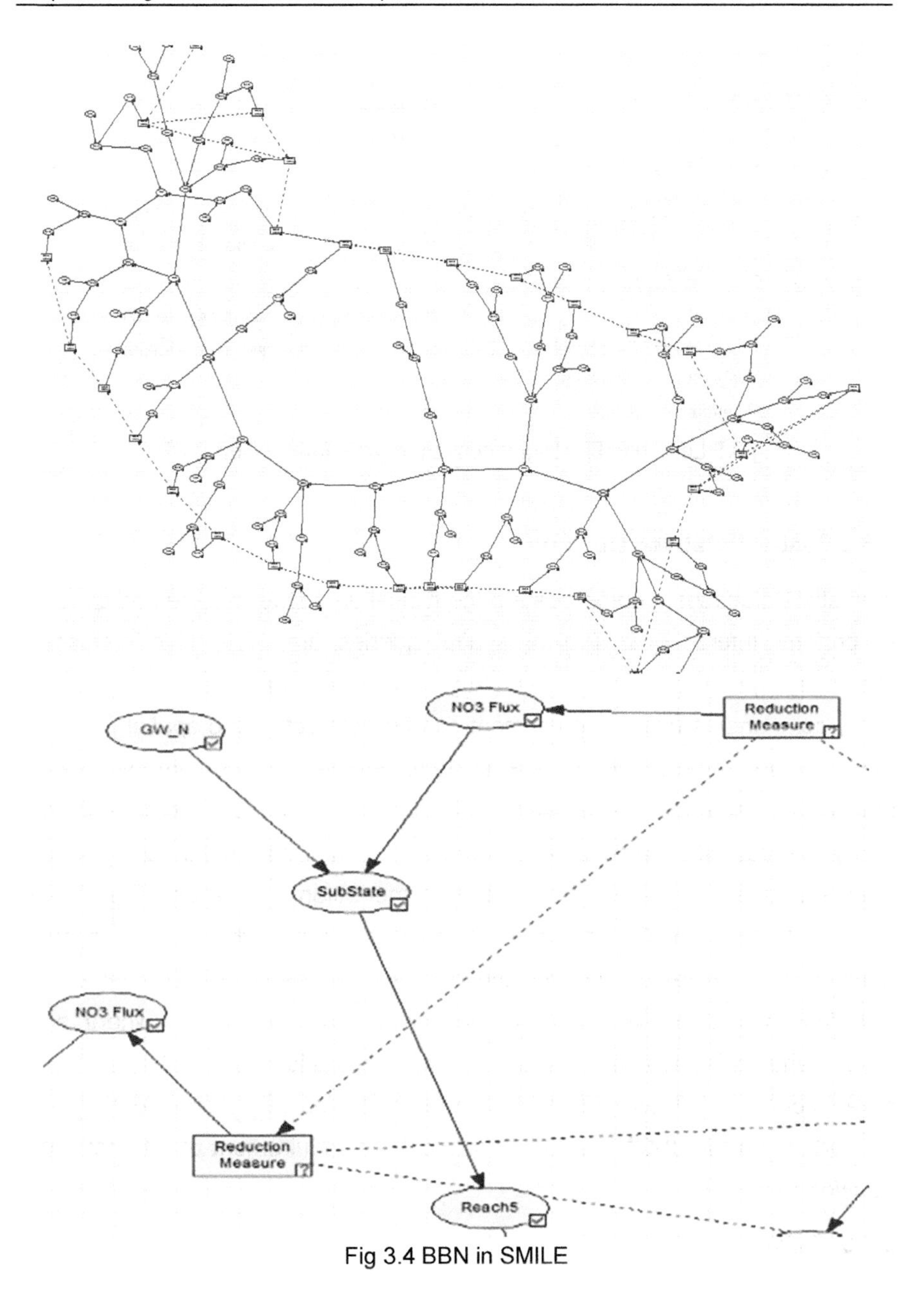

Fig 3.4 BBN in SMILE

```
node SubState_6 {
    .......
    parents = (NO3_6, GW_N6);
    states = (Good, Bad);
    probs =
        (((0.20000000,
0.80000000),
        (0.35000000, 0.65000000)),
        ((0.35000000, 0.65000000),
        (0.80000000,
```

```
11, 0.20000000, 0.80000000, 0.35000000, 0.65000000,
0.35000000, 0.65000000, 0.80000000, 0.20000000

6, 0.20000000, 0.80000000, 0.35000000, 0.65000000,
0.35000000, 0.65000000, 0.80000000, 0.20000000

8, 0.20000000, 0.80000000, 0.35000000, 0.65000000,
0.35000000, 0.65000000, 0.80000000, 0.20000000
```

Fig 3.5 Left: .dne file; Right: converted table format file for CPT

3.3.2 Spatial Database Design

In spatial databases, geometries are the typical data type to store and queries to perform interactions. Typical spatial queries are: determine distances (geometry, geometry), areas (geometry) and so on. In water management, many components inside a catchment can be abstracted into spatial features that can be entered into a GIS system. Shape files contain geospatial primitives to represent the geometry of the objects. A geospatial database can easily interact with shape files for retrieving spatial information and extend their attributes. In the Vis-Reacher implementation, the shape files were converted into a PostGIS database. The CPTs maintain their spatial identity number for each feature so that these tables can be linked together using SQL functions. Generally speaking, the table contains the 'what',' where' and 'how'. 'What' is the type of the feature, 'where' is the location, and 'how' is the parent node that influences this node. After the tables are completed, they are converted into the JSON file format to meet the requirement of Internet data transformation.

3.3.3 VisREACHER Design

The VisREACHER is the visualisation module of the REACHER tool designed to be one single webpage. Its functionalities are to interact with the spatial database and present the BBN computational results inside the browser in order to execute the BBN model directly, based on the users' input. The well-known Client/Server architecture is used as the main framework with the

Browser as client. Several FOSS techniques were selected to build the system, as discussed hereafter.

Server part:

Apache: is considered the most popular open source HTTP server software. The primary use of this software is to serve both static and dynamic web pages. It supports a variety of features and language interfaces, e.g. PHP, Perl and Python.

Mapserver: is a graphic data rendering engine at the server side. In its most basic form, it is a CGI (common gate interface) program that sits inactive on a Web server until a request is sent to the Mapfile to create an image of the requested map. The request may also return images for legends, scale bars, reference maps, and values passed as CGI variables. The mapfile is the heart of the mapserver, it defines the relationships between different objects, points Mapserver to where data are located and defines how to drawn the map (http://mapserver.org/introduction.html). One advantage of 'Mapfile' is that it can link to the spatial databases directly. In this project, we use Mapfile to create our own static map layers for the whole project area.

Client part:

Ajax: is not a single technique but a group of techniques to enhance the Internet users' experience. In the traditional web page, the users need to refresh the page very often to get new data, wasting time in waiting for response. The Ajax technique is to use an asynchronous way to request parts of the data update in the server so that not the whole page will be refreshed. The feeling for the user is that they get the new data very fast and need not be facing blank pages all the time. A mature example of using Ajax is Google Maps. The main techniques in Ajax include: JavaScript, XMLHttpRequest, Cascading Style Sheet (CSS) and Document Object Mode (DOM).

Openlayers: is a JavaScript library to help generate the interactive user interface between maps and layers. As of November 2007, OpenLayers has become an Open Source Geospatial Foundation project (OSGeo). It has no-server side dependency so that it can support the mapserver to render the user defined maps.

Extjs: is a cross-browser JavaScript library for building rich Internet applications. Mainly, it adds more complex options tools to the user interface, e.g. menus, forms, tabs, and so on. They are used here for arranging the layout about VisREACHER.

WebKit: is an Open Source web browser engine which supports CSS 3D transforms needed to build the 3D Popups for the applications.

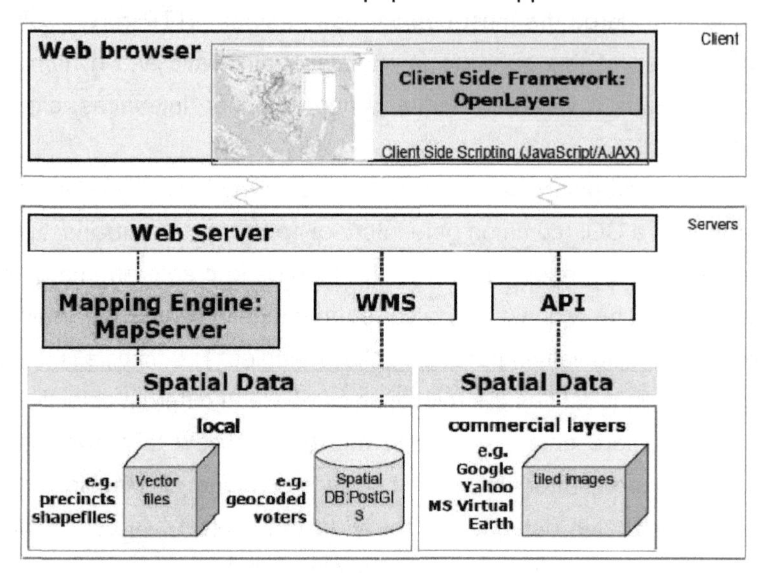

Fig 3.6 Stucuture of VisREACHER

3.3.4 Interaction Design

Interaction Design is a relative new discipline since (Moggridge and Verplank, 1980) first coined the term. It addresses the improvement of the software interface without giving concern to how the underlying functionalities are working. The topic deals with the human side of Human-Machine interaction and the methodologies are related to cognitive research, which focuses on human behaviour and interaction. In this research the interaction is based on maps. Map-based systems are familiar to many end-users due to the prevalence of GIS, Google Maps and Google Earth. The essential reason why so many people like to use or at least have a try on Google maps products is that it is easy to use. Hence this approach was used for Vis-REACHER design. Based on a survey on preferred interactive styles for map-based systems, the following features were incorporated:

- Layer control
- Clickable Items
- Tool bars

- State representations
- Interactive Devices

Vis-REACHER uses maps as the main window and layer control for the different information patterns. Clickable features with popup menus works for representing various measures and provide triggers to display the changes on the specific layer. 3D popups are introduced in this research to display different information components simultaneously. By rotating, users can find their specific measures and select options. All modifications of the measures are explicitly displayed on the 3D popup.

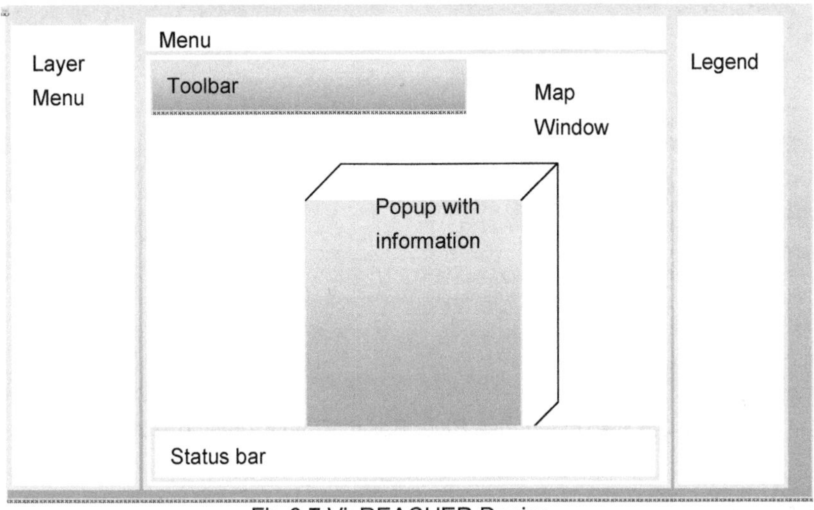

Fig 3.7 VisREACHER Design

3.4 Implement REACHER in Odense River Basin

3.4.1 Description of study area

The Odense river basin is located on the island of Funen that is the second largest Danish island. The whole area is around 1050 km2 which includes 1015km of water streams. The Odense river, which is about 60 km long and drains a catchment of 625 km2, is the largest river in this area. The population of the river basin is 246,000 and 10% of the population is not serviced by sewerage systems. The monthly precipitation ranges from 40mm (April) to 90mm (December/January). Clay and sand are the major soil types in this river basin. There are four land use types: farmland (68%), urban area (16%), woodland (10%) and natural/semi-natural areas (6%). Obviously, agriculture is dominant in this area and there are 1870 registered farms in the database. Some 960 of these farms are livestock holdings and the livestock density is about 0.9 unit/ha. Therefore, the main pollutants are Nitrate and Pesticides. The SWAT model developed for this study area computes Nitrate and Pesticide pollution to the river Odense and evaluates the effectiveness of different measures to reduce pollution to the river (Hoang, 2013).

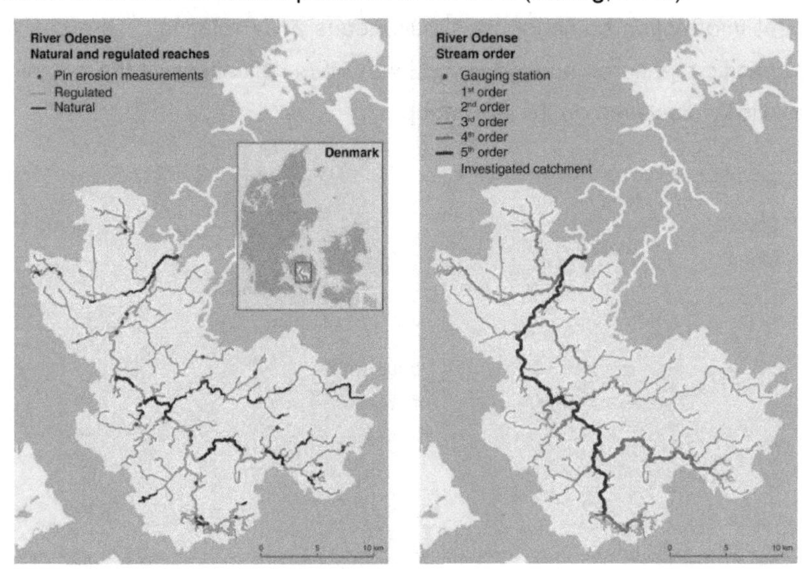

Fig 3.8 Map of Odense River Basin (Brian Kronvang, 2010)

3.4.2 Input data

In this research, thirty sub basins were identified in the study area and the topological information about the study area was contained in the shape files, from which the upstream/downstream relations can easily be determined. The SWAT model simulations provide nine years of daily simulation results about nitrogen, pesticides and more. The SWAT modelling tool uses files for database storage (see sample in Fig. 3.9).

```
General Input/Output section (file.cio): ArcSWAT 2.1.6
14/07/2010 00:00:00ARCGIS-SWAT interface AV
```

Year	Day	Hr	FLOWm^3/s	SEDmg/L	ORGNmg/L	ORGPmg/L	NO3mg/L	NH3mg/L	NO2mg/L	MINPmg/L	CBODmg/L
1990	1	00	0.000E+00	0.000E+00	0.000E+00	0.000E+00	0.000E+00	0.000E+00	0.000E+00	0.000E+00	0.000E+00
1990	2	00	0.108E+00	0.000E+00	0.567E+00	0.170E-03	0.317E+01	0.686E+00	0.631E-04	0.429E-03	0.000E+00
1990	3	00	0.428E-01	0.000E+00	0.635E+00	0.763E-04	0.154E+01	0.467E+00	0.410E-01	0.363E-03	0.000E+00
1990	4	00	0.117E+01	0.000E+00	0.852E+00	0.998E-04	0.217E+01	0.591E+00	0.125E+00	0.663E-03	0.000E+00
1990	5	00	0.977E+00	0.000E+00	0.119E+01	0.152E-03	0.292E+01	0.860E+00	0.903E-01	0.113E-02	0.000E+00
1990	6	00	0.891E+00	0.000E+00	0.114E+01	0.148E-03	0.300E+01	0.868E+00	0.102E+00	0.130E-02	0.000E+00
1990	7	00	0.859E+00	0.000E+00	0.112E+01	0.149E-03	0.306E+01	0.867E+00	0.106E+00	0.144E-02	0.000E+00
1990	8	00	0.851E+00	0.000E+00	0.109E+01	0.144E-03	0.311E+01	0.870E+00	0.115E+00	0.156E-02	0.000E+00
1990	9	00	0.854E+00	0.000E+00	0.107E+01	0.141E-03	0.314E+01	0.870E+00	0.121E+00	0.165E-02	0.000E+00
1990	10	00	0.860E+00	0.000E+00	0.105E+01	0.139E-03	0.317E+01	0.871E+00	0.126E+00	0.171E-02	0.000E+00
1990	11	00	0.866E+00	0.000E+00	0.102E+01	0.129E-03	0.318E+01	0.876E+00	0.140E+00	0.176E-02	0.000E+00
1990	12	00	0.871E+00	0.000E+00	0.992E+00	0.123E-03	0.319E+01	0.878E+00	0.150E+00	0.180E-02	0.000E+00
1990	13	00	0.875E+00	0.000E+00	0.102E+01	0.133E-03	0.319E+01	0.871E+00	0.135E+00	0.178E-02	0.000E+00
1990	14	00	0.878E+00	0.000E+00	0.105E+01	0.142E-03	0.318E+01	0.865E+00	0.124E+00	0.177E-02	0.000E+00
1990	15	00	0.880E+00	0.000E+00	0.107E+01	0.150E-03	0.318E+01	0.858E+00	0.114E+00	0.175E-02	0.000E+00

Fig 3.9 Sample SWAT model results

3.4.3 Various Layer Types

Mapserver was chosen as the rendering engine for this purpose and mapfile was used for conversion into Web Map Server (WMS) layers in the OpenLayers configuration and presented as static layers for displaying the general conditions of the study area. Vector layers were used as dynamic layers that link to the JSON file directly though the OpenLayers' HTTP protocol to get the spatial features, using the style() function to display the features in different colour. One function named FeatureStore in the GeoExt library was used here to generate a temporal cone file from the original JSON file in order to support the modification functions, i.c. the BBN core engine. The advantage of this approach is that the baseline of the scenario is fixed so that the starting scenario will not be changed when refreshing and opening links on the webpage, unless users update or save the baseline file with new scenarios manually (see Fig. 3.10).

```
MAP
   NAME map_generated_by_gvsig
   SYMBOLSET "OB1.sym"
   SYMBOL
     NAME "arrow"
     TYPE TRUETYPE
     FONT "vera"
     CHARACTER ">"
     GAP -15
END
   FONTSET "fonts.txt"
```

```
var BBN = new OpenLayers.Layer.Vector("BBNNO3", {
    strategies: [new OpenLayers.Strategy.Fixed()],
    protocol: new OpenLayers.Protocol.HTTP({
        url: "data/BBN",
        format: new OpenLayers.Format.GeoJSON()
    }),
    styleMap: new OpenLayers.StyleMap(new OpenLayers.:
});
```

Fig 3.10 Left: Sample Mapfile; Right: Vector Layer Declaration in OpenLayers

3.4.4 Client-side BBN configurations

The BBN model contains basically two components: the graphic network and the conditional probability table. The graphic network is contained in .shp files which hold the topological relationships between upstream and downstream. The BBN network is built in SMILE with the output .dne file. The conditional probability table can be abstracted from the .dne file into the database format and converted into the JSON file format. With the BBN simulation engine inside the client-side, VisREACHER has the ability to reproduce the computing scenarios for the entire network. The main task of the engine is to calculate the various probabilities propagating through the BBN graph network. In Odense river basin, four graph structures were defined related to the fundamental processes.

First is the basic structure inside the same sub basin:

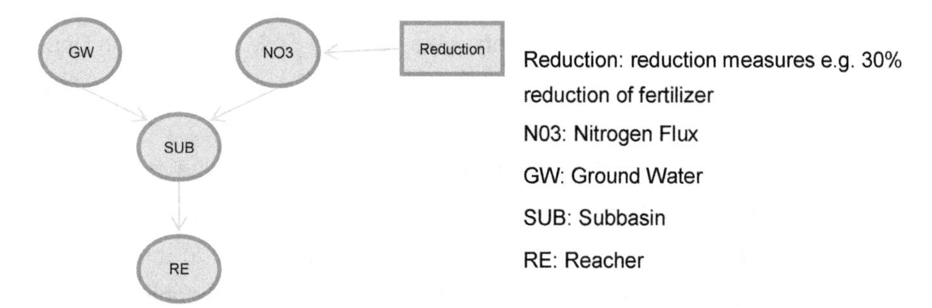

Fig 3.11 Structure in the same subbasin

This graph structure illustrates the relationship in the sub basin on how the reduction factor influences the reaches. Therefore, there are three CPT tables in this structure: between Reduction and NO3, between GW-NO3 and SUB, between SUB and RE. The relationship between reduction measures and nitrogen change is calculated by Monte Carlo simulation with the original SWAT model to estimate the effect of measures, e.g. a 30% reduction of fertilizer.

There is a clear relationship from upstream to downstream, where three situations can be distinguished: (there is an assumption that one downstream can only has two upstream at most)

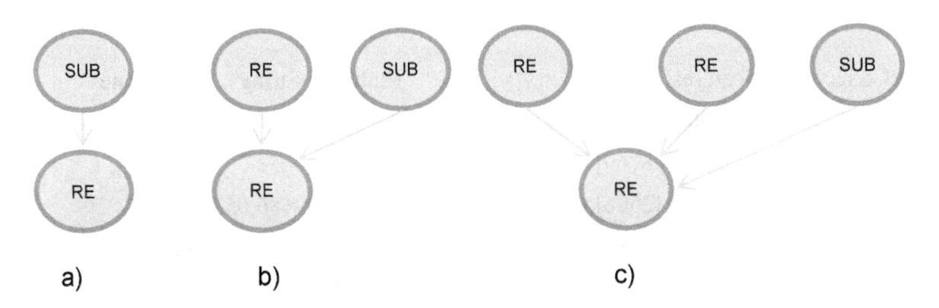

Fig 3.12 a) means it is the first upstream in the river basin or head node in the graphic structure; b) means it has one upstream or one father structure; c) means it has two upstream or parent structure.

The spatial database is designed to couple the spatial information with the CPTs and has been converted to JSON format for Internet transformation.

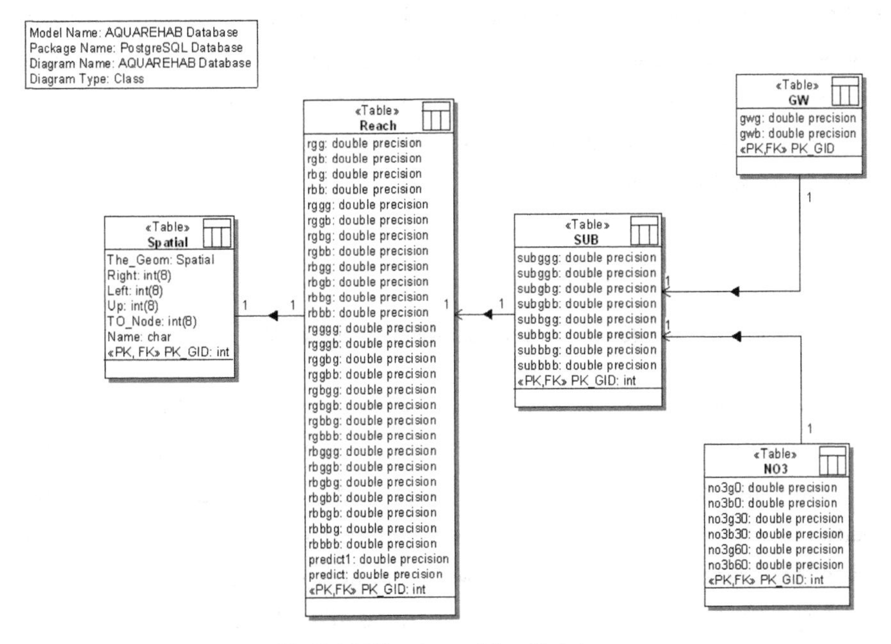

Fig 3.13 Sturcture of the Database

{"type": "FeatureCollection", "features": [{"geometry": {"type": "MultiLineString", "coordinates": [[[1160334.7350028532, 7439902.0052383784], [1161037.5833003318, 7439887.5657335902], [1161209.6809750944, 7439707.8545568269], [1161568.3394556844, 7440052.7890927307], [1161919.7691329715, 7440045.5210442506], [1162263.9222211039, 7439686.052856477], [1162260.2846408142, 7439509.9667282384], [1162432.3454623451, 7439330.2373986086], [1163310.8333778337, 7439311.9416788854], [1163490.1955903508, 7439484.3512846138], [1163841.596869353, 7439476.9966756385], [1164020.9745952408, 7439649.3974592118]]]}, "id": 2, "type": "Feature", "bbox": [1160334.7350028532, 7439311.9416788854, 1164020.9745952408,

Fig 3.14 Content of JSON file

Accordingly, there are twenty-seven probability combinations over the two states of the reaches. With these probabilities, the function Total Law of Probability was used to calculate the conditions in the reaches. The reason why is that the purpose was to know how 'good' the situation in the reaches will be. Therefore, the model only considers the probabilities of a 'good' state in each reach so that the model needs to sum of total possibility for achieving such 'good' state, according to:

$$P(A) = \sum_{n} P(A \mid B_n) P(B_n)$$

Equ.3.1. Total Possibility Calculation

where 'A' represents the 'Good' states; Bn represents the condition in reach n; and $P(B_1)$ represents the probability of **'Good'** states. Then, $P(A \mid B_n)P(B_n)$ represents the possibility of a downstream reach being 'Good' when the upstream reach n is 'Good'. The situation can also include the possibility of a downstream reach becoming 'Good' when the upstream reach is 'Bad'. After summing all possible combinations between the effect factors of the upstream and downstream relationship, the total value of the probability of the specific reach 'Good' states can be represented.

In the Odense case, there were two kinds of reduction measures considered for fertilizer: 30% reduction and 60% reduction, relativey to the No Change (NC) scenario. NC is very important for representing the other unmodified reach states. Meanwhile, the initial states of reaches need to be identified under the NC scenario for fertilizer. The flowchart of the BBN simulation core engine in the Javascript program is shown in Fig.4.5. The process of calculating probabilities follows Eq. 4.1. The value of different probabilities for $P(A \mid B_n)$ and $P(B_n)$ is read from the JSON file, which is created from the SIMLE .dne file. After this, the probabilities can be propagated though the network from upstream to downstream which is built upon the river basin's topology. Therefore, users can immediately view the changes over the entire river basin and combine different measures over different spatially distributed features.

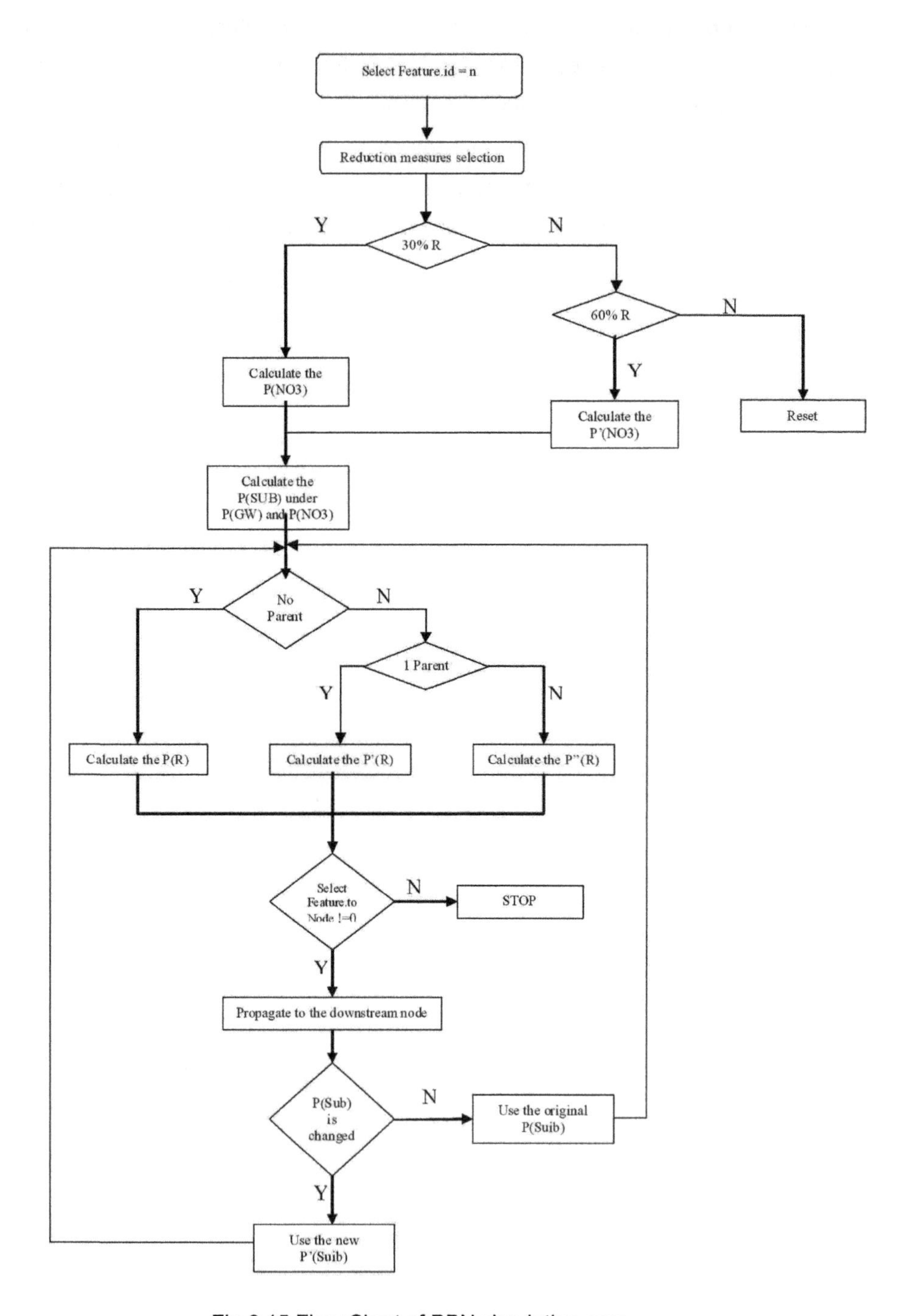

Fig 3.15 Flow Chart of BBN simulation core

3.4.5 Extend Measures Selection from 2D to 3D

Clickable items are considered as the most important feature in this web-based DSS. The 2D popups were covert into 3D in order to reduce the users' movement among the web browsing activities and also extend the dimension for options. Two items were addressed in this thesis research:

- How to build the 3D popups?
- How to control them?

The methodology of drawing 3D popups depends on the support from the rendering engine for the web pages. WebKit was chosen to be the layout engine for the VisREACHER development. Chrome and Safari are used as the test browser environment due to their natural relationship with WebKit. The main reason to select WebKit is that it supports the Cascading Style Sheets 3D (CSS 3D) transformation since 2007. Different from the other web 3D solutions, e.g. Java 3D, Viewpoint, X3D and so on, WebKit does not need to install any plug-ins and is real-time rendering on the client side. The Webkit-transform property was used to specify the transform over a group of CSS elements. In CSS, the 3D object has been divided into several elements, which can be represented in 2D and then projected in a three-dimensional space having viewport with Z-axis. Then, readily available functions can be used to realize the transformations over the 3D object, e.g. rotate, scale and translate. Although this increases the complexity of building 3D models and may not have many state-of-art rendering techniques used for building a fancy scene, it is enough to generate fundamental 3D objects to merge inside a webpage for extending the user experience. Especially, when users not require a fully dynamical virtual environment, a combination of 2D and 3D applications is an adequate choice to design the user interface. Here, six faces have been created to compose a 3D cube for the popup by specifying the size and position of each square. The contents on the different faces in contained in HTML pages that link with the main map-based window. The question about how to control the 3D popup comes back to the Javascript environment. The VisREACHER is using a nested document element model (DOM) structure and the birth of the popup depends on the user selection. A click on the specific element can be considered as an 'event' which will trigger execution.

3.5 Evaluations

3.5.1 Overview

The overview of the VisREACHER is shown in Fig 3.18. The main page is separated into four parts: map window, tool bar, legend bar and status information. The layer control can be selected from the tool bar and the box can be dragged anywhere onto the screen which solves the problem of matching different screen resolutions and saves space for the viewport. The Layer Control contains two classes: Base Layers and Active Layers. The first one mainly represents the basic topological information and position of the spatial features. The Active Layers can be changed dynamically by editing the parameters.

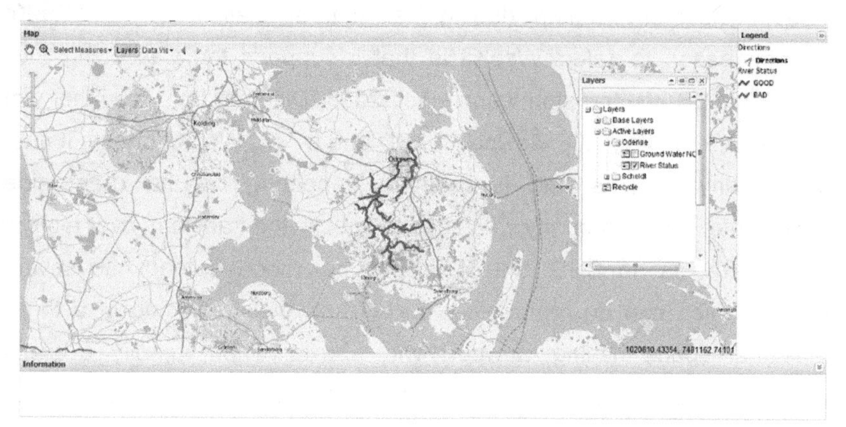

Fig 3.18 Overview of the VisREACHER

3.5.2 Exploring Effects of Remedial Measures

Remedial measures are the most important component in the REACHER tool, helping stakeholders understand what are the consequences of possible measures. Because agriculture is the dominant industry in the study area, reduction of fertilizer has been chosen as the first remedial measure. The assumption is that the reduction measures are working in particular reaches so that the user can click on the reach for the selection of the measures. For simplicity only 'Good' or 'Bad' are considered here, using Green and Red to represent these states for obtaining a clear view of the effect of measures. Two measures are prepared to achieve the 'Good' scenarios, i.e. 30% reduction in fertilizer, and a 60% reduction in fertilizer. After clicking on the layer 'River Status' in the 'Active Layers' menu and select the selection mode from the 'select measures' option, users can click on the reach they want to implement the measures. They can single out reaches or press 'Ctrl' and then group select different reaches. After clicking on the specific reach, a popup menu will show the options about the measures and the information about the selected items. Then, the user can select the measures for those reaches. The selection will trigger the simulation core to calculate the update scenarios for the river basin and propagate the results to the downstream area. The status bar will record what has been done till this moment. In the pilot study, we made an assumption to link the measures over the reach to the groundwater polygons to test the ability of such modification. In this test, users operate the remedial measure on the reach and then view the impact on the groundwater. In real cases, the functions of such linkage need to be identified. Although the functions are assumed now, the method to visualize this cross-layer impact is the same and remains flexible to add new functions. The Figures below show the effects of simulated measures.

Fig 3.19 Overview of the VisREACHER

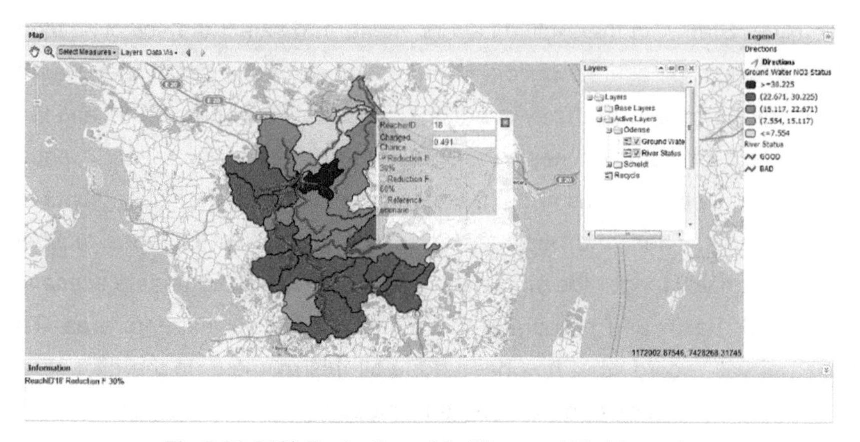

Fig 3.20 30% Reduction of fertilizer on NO.18 reach

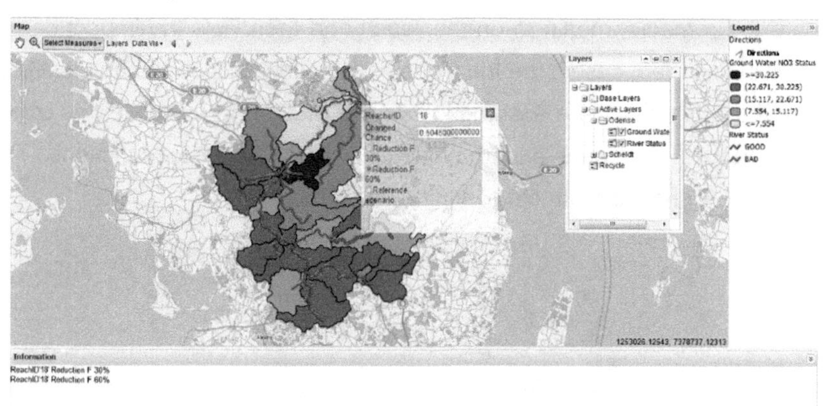

Fig 3.21 60% Reduction of fertilizer on NO.18 reach

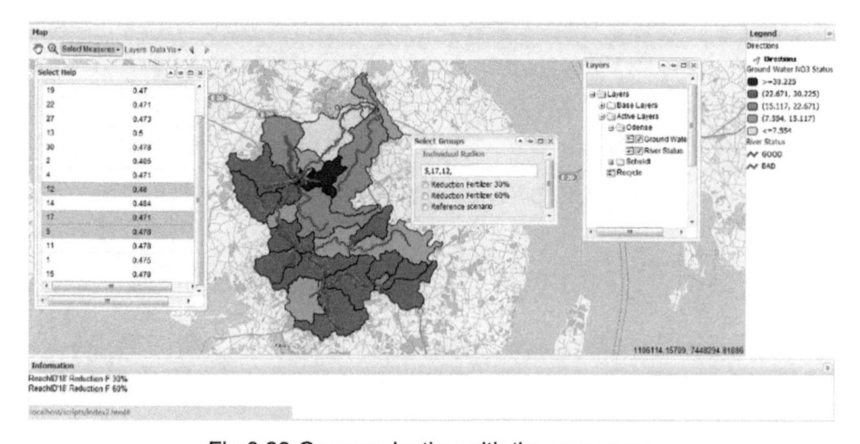

Fig 3.22 Group selection with the measures

3.5.3 3D Popup Performance

The process to active the 3D popup menu is the same as described in the previous section. The tail version of this 3D popup is tested on the Safari (mobile version) and Chrome (version 15.0) browser. In the tests, the 3D object can properly be rotated. Each side can link to a different HTML page for more information about the selected reach. In this way, the information dimension can be extended for the user. The rotation function can only be activated when the left button of the mouse is pressed so that the select function inside the HTML page can work properly after the left button is released. Therefore, the control of the rotation and selection can be done the mouse with one finger press movement. The 3D popup menu works well under the mouse device whereas the option to select items works better in a touch screen device environment.

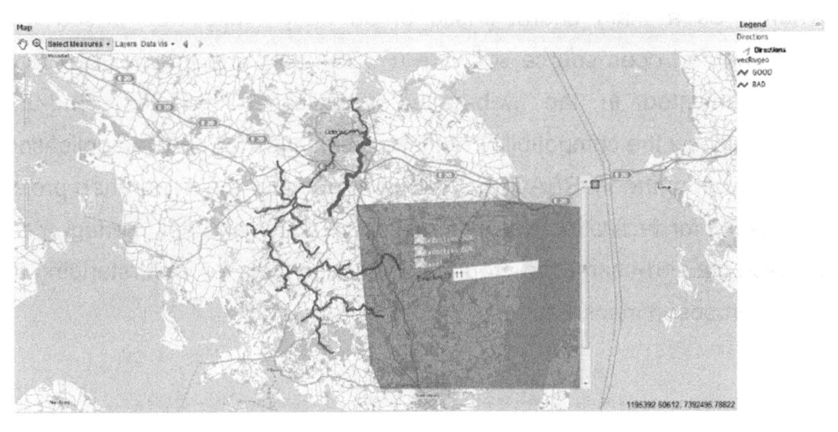

Fig 3.23 3D popup with the measures

3.6 Summary

In decision support processes, groups of users include different stakeholders from different areas. The requirements from the users not only focus on a specific modelling environment to predict future scenarios, but also on easy-to-use tools for them to understand the impact of proposed measures. Modelling and visualisation are two important components in the REACHER tool design. In order to achieve the necessary performance speedup, the process-based SWAT model was approximated by a 'surrogate' Bayesian Belief Network (BBN) model, in order to reproduce the essential characteristics of the more complicated model. The main motivation for building the surrogate model are runtime saving and flexibility to extend different measures at the same time. The challenge of any easy-to-use tool is to find an appropriate visualisation method to interact with the end-users between the UI and the surrogate model. Map-based user interfaces can give clear information on the spatial characteristics by combining abstract items with a realistic map. Clickable items and Layers Control are the most important factors in map-based GUI design. By clicking on the spatial features on the map, virtual measures can be implemented. Upstream and downstream relationships are represented explicitly that can help the user understand more about the impact of measures on the whole river basin. 3D popup menus can extend the dimensions of option selection in order to maintain the clear layout and save screen space for the GUI. Ajax techniques are important components in Web 2.0 that help create smooth user experience as a main characteristic of an easy-to-use web based tool. Several free and open source software (FOSS) tools and Internet standards were implemented in the web-based visualisation modules. End-user evaluations show the compatibility of those techniques for on-line applications. The framework of the VisREACHER tool developed for the Aquarehab project proved useful for facilitating decision making in water and soil management, and served the end-users who require easy-to-use tools for understanding the impact of proposed measures.

4 Using 3D Virtual Environments for Displaying Effects of Land use Change

Case Study for lower Yellow River Delta

4.1 Introduction

At the Yellow River Conservancy Commission in China, the 'Digital Yellow River' concept was developed for better managing the Yellow River System (Digital River Hand Book, 2006). This project combines Remote Sensing data, Geographical Information Systems, Global Positioning Systems, and numerical simulation modelling systems to construct a virtual platform that is able to visualize the effect of possible developments and management scenarios to decision makers. In that sense, the 'Digital Yellow River' is the contraposition of the physical Yellow River in a virtual computer-based environment. Such environment needs 3D (or even 4D) advanced visualisation features. The system should be able to combine 3D graphics techniques with 2D GIS maps as well as with results from computer-based inundation modelling, in order to provide decision makers with the visual support environment they need. In fact, it already proves quite helpful if stakeholders (in the case of the Yellow River the Commissioners and their advisors) can assess the effects of proposed flood prevention measures directly in geographic setting by interaction with the virtual environment. In this way decision makers can e.g. view the location of particular hydraulic structures like reservoirs and dams, while the actual water level in the river is represented as a transparent texture file. There are two fundamental features in a virtual river environment: the surrounding terrain map and the actual water surface. Underground features like groundwater tables may also be of relevance, but are not considered in this research. In this chapter, the digital terrain representation is the main consideration as elaborated in the following sections.

4.2 Digital Terrain Simulation

Generally speaking, the digital terrain simulation process consists of building a 3D terrain map from elevation data measured at a preferably large number of locations. The constructed terrain maps can easily be displayed on computer screens while the viewpoints and scales can be easily adjusted. Texture mapping can be used to create photo-realistic views of the digital information. There are two commonly used formats for DEM data, Raster and TIN, which can be converted into each other.

The process of acquiring DEM data is not the main topic in this research, so the data used in this thesis is already interpolated from the sampling points. DEM data can be transformed into an Elevation Map that can be stored as a raster image. In each pixel the real elevation value is mapped onto a chromatogram, e.g. using a grey shades to store elevations varying between values ranging from 0 to 256. Elevation Maps contain the basic information about the terrain elevation and can be used to display 3D terrain features using computer graphics techniques. The Open Graphics Library OpenGL (Dave et al., 2007) is a standard specification defining a cross-language, cross-platform API for writing applications that produce 3D (and 2D) images. The interface consists of over 250 different function calls that can be used to draw complex three-dimensional scenes from simple primitives. OpenGL was developed by Silicon Graphics and is popular in the video games industry. OpenGL is widely used in CAD, virtual reality simulation, scientific visualisation, information visualisation, flight simulation and video game development. In this research mainly the functions inside the graphic library were used to realize 3D images of the Yellow River delta. Object-oriented programming was chosen to implement the system due to the advantage of easily abstracting different classes from the real data and memory saving (Bruce 1995).

Research on algorithms that can reduce digital terrain information has been receiving considerable attention. The Level of Detail (LoD) technique as outlined by (Shaffer et al., 1993), contains two main principles: one based on quadtree partitioning and one based on bintree partitioning. Hugues (Hugues 1998) proposed a Progressive Mesh model, which uses a bottom-up mode to subdivide the terrain nodes. He used an unstructured simplification mode to add/reduce details contained in an arbitrary set of connected triangles. Although very attractive from a computer science point of view, this method seems too complex to be implemented in simplified practical applications. Lindstrom (Lindstrom et al., 1996) and Koller (Koller et al., 1995) used a method based on the quadtree data structure to partition the terrain into many tessellates, then build an approximate elevation map. We choose the quadtree as our data structure for the terrain data due to the simplification. Röttger (Röttger, Heidrich et al. 1998) gave a very efficient node evaluation system for judging whether or not a terrain node needs to be subdivided. In section 4.4, we will use this idea as the reference for introducing our node evaluation system and give the implementation results for flood plain simulation.

4.3 3D Terrain Simulation

4.3.1 3D Skeleton construction

Composing the skeleton for the 3D terrain model is of course creating an abstraction of the real world topography. Based on actual values sampled during topographic surveys at selected points, a continuous surface is constructed using an interpolation approach similar to the finite element method in computational sciences. Such approximated terrain map can be entered into a database system, which then makes it possible to select and explore various views on the data (i.e. terrain map), making use of the same basic concepts linking objects, attributes, relationships and operations (Tsichritzis, 1982). Like in any abstraction, there can be simple objects like 0D, 1D, 2D, 3D representations, and complex objects as combinations of them. If we abstract the primitives from 3D models, two kinds of primitives can be obtained: surface-based models and volume-based models. The first one focuses on the outside surface of the model (in our case the landscape terrain model) and the second takes into account the interior as well. From the point of view of Roberto (Roberto, 2006), the different methods for describing those two classifications are shown in Figure 4.1. Voxels are 3D volume pixels used for representing a solid by density functions. The advantage of this method is that the volume of the objects can easily be calculated, with the disadvantage that the surfaces are always rough. In this research, the riverbed terrain surface is the main object we need to simulate so that the surface model is more relevant here.

4.3.2 Surface Modelling

There are three common methods for surface modelling: grid-based, triangle-based and NURBS-based methods. The first two methods are interchangeable and the last one is good at simulating free-form surfaces. For terrain representation, DEM data are the most common basiss. Raster (regular grid) and TIN (triangular irregular network) are two formats for DEM. Grid-based models and triangle-based models are both suitable for 3D terrain representation. Quadtree-based triangulation is the most efficient method in 3D terrain modelling, provided the input data has a regular structure. The elevation map is the main focus in this research and can be derived from the 3D terrain data in either data format. The next sections will introduce the main algorithms for surface display, transformation and information query.

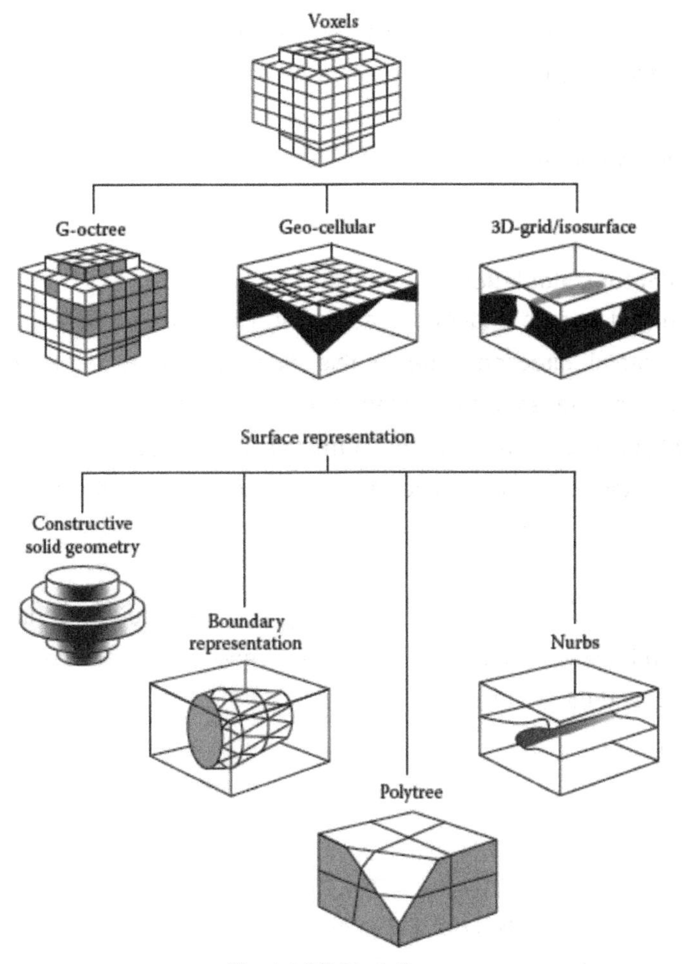

Fig 4.1 3D Modeling

As mentioned in section 4.2, the Elevation Map is used to store the elevation information of the terrain. Quadtree-based triangulation is used to present this map in a 3D setting, creating a virtual 'feel' for the third dimension. From the elevation map, the 3D terrain can be obtained by taking the positions of the elevation vertices and linking them to the edges in the identified 3D space. However, this only reveals the geometry information about the terrain. Without the appropriate 'shallow-view' representation, the simulated terrain looks like a thematic object in a visualisation system. So a mapping technique is used to add the realistic feeling of the third dimension.

Bump mapping is a method to simulate bumps and winkles on the surface of an object, which is very suitable for digital terrain modelling. In bump mapping, the shading function depends on the roughness of the texture file and the elevation map can be used as a kind of texture file. Bump mapping is very useful when the normal vector is changing in accordance with the roughness.

Normal Vector Calculation

The DEM grid can be understood as having lots of little different faces, with different normal direction. Due to this, light will be reflected in different directions. So, the software system should determine the direction of the reflected light. The definition of the normal vector can be used for solving this problem. In this system, we use the triangle to calculate the normal vector. The methodology is described below.

Given three 3-vectors P_1, P_2, P_3 defining the three point's vertices of the triangle and using elementary vector subtraction, the two edges of the triangle, E_1 and E_2, become

$$E_1 = P_3 - P_2 \quad (4.1)$$

$$E_2 = P_2 - P_1 \quad (4.2)$$

The definition of the surface normal vector is a unit-length vector that is perpendicular to any vector in the plane of the surface. By taking the cross-product of the two vectors E_1 and E_2 in the plane of the surface, Equation 4.3 gives the un-normalised surface normal vector

$$n = E_1 \times E_2 \quad (4.3)$$

which can be normalised like like any vector v, taking $|v| = \sqrt{v_x^2 + v_y^2 + v_z^2}$ so that

$$\hat{n} = \frac{1}{|n|} n \quad (4.4)$$

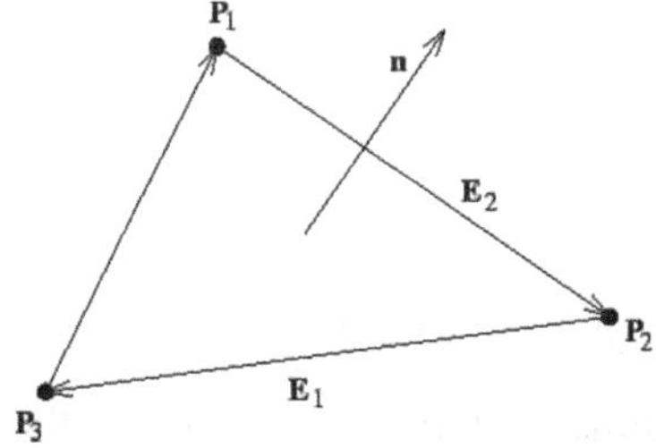

Fig 4.2 Geometry of the given triangle and its surface normal

If the points P_1, P_2 and P_3 are defined in a clockwise order and the surface normal is required to be in the "upward" direction as depicted in Figure 4.2, then the vectors E_1, E_2 and $\hat{n}$ form a right-handed set, as is usually desired.

Combining Equations 4.1, 4.2 and 4.3 we obtain Equation 4.5.

$$n = (P_3 - P_2) \times (P_2 - P_1) \tag{4.5}$$

Expanding this into the individual components, and using the definition of vector cross-product, we obtain the three Equations 4.6 to 4.8.

$$n_x = (P_{3y} - P_{2y})(P_{2z} - P_{1z}) - (P_{3z} - P_{2z})(P_{2z} - P_{1y}) \tag{4.6}$$

$$n_y = (P_{3z} - P_{2z})(P_{2x} - P_{1x}) - (P_{3x} - P_{2x})(P_{2z} - P_{1z}) \tag{4.7}$$

$$n_z = (P_{3x} - P_{2x})(P_{2y} - P_{1y}) - (P_{3y} - P_{2y})(P_{2x} - P_{1x}) \tag{4.8}$$

In order to normalise these values, we find the scalar value n, using the values for n_x, n_y and n_z found in Equations 4.6 to 4.8.

$$n = |n| = \sqrt{n_x^{\ 2} + n_y^{\ 2} + n_z^{\ 2}} \tag{4.9}$$

to find the normalised versions of Equations 4.6 to 4.8 as follows

$$\hat{n}_x = \frac{1}{n} n_x = \frac{1}{n}(P_{3y} - P_{2y})(P_{2z} - P_{1z}) - \frac{1}{n}(P_{3z} - P_{2z})(P_{2y} - P_{1y}) \tag{4.10}$$

$$\hat{n}_y = \frac{1}{n} n_y = \frac{1}{n}(P_{3z} - P_{2z})(P_{2x} - P_{1x}) - \frac{1}{n}(P_{3x} - P_{2x})(P_{2z} - P_{1z}) \tag{4.11}$$

$$\hat{n}_z = \frac{1}{n} n_z = \frac{1}{n}(P_{3x} - P_{2x})(P_{2y} - P_{1y}) - \frac{1}{n}(P_{3y} - P_{2y})(P_{2x} - P_{1x}) \tag{4.12}$$

The values $\hat{n}_x$, $\hat{n}_y$ and $\hat{n}_z$ given in Equations 4.10 to 4.12 are now the x, y and z coordinates of the normalised surface normal to the triangle with vertices $P_1 = (P_{1x}, P_{1y}, P_{1z})$, $P_2 = (P_{2x}, P_{2y}, P_{2z})$ and. $P_3 = (P_{3x}, P_{3y}, P_{3z})$

Average Normal Vector Calculation

The earth surface can be represented by a sphere that can be approximated by a large number of plane triangles fitted to the curved surface. Using the normal vector calculation, we can obtain the normal vector for each triangle (Fig 4.3).

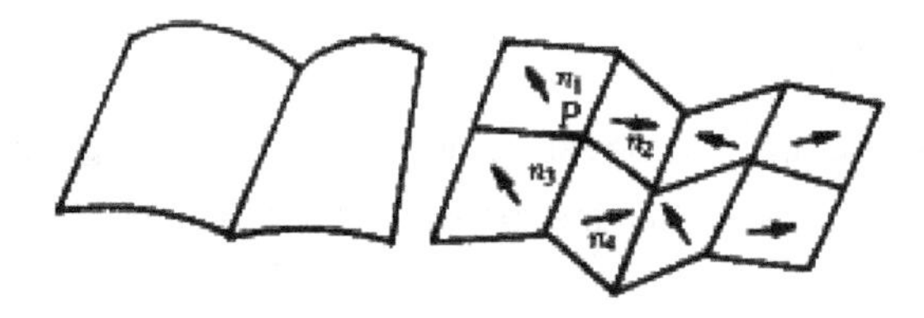

Fig 4.3 Normal Vector

However, if the whole curved surface is approximated by using this algorithm, the result will be not smooth under light reflection and the edges and corners become very obvious. Therefore, the average normal vector calculation is needed. In the system developed here, the average normal vector calculation is carried out for 3 aspects: the grid point on the edges (not included the boundaries), the 4 corner points, and the 4 edges.

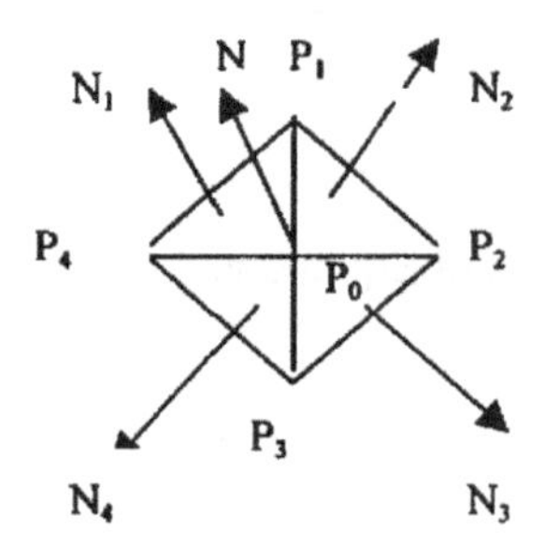

Fig 4.4 Average Normal Vector

The normal vector of the 4 surfaces $(P_0P_1P_4, P_0P_1P_2, P_0P_2P_3, P_0P_3P_4)$,which surrounding the P_0 is N_1, N_2, N_3, N_4 , so the P_0's average normal vector is $N = (N_1 + N_2 + N_3 + N_4)/4$

Coordinate Transformations

Graphic Transformation is another core algorithm for 3D graphics because it relates to the interaction with the 3D models. The transformation includes three parts: transformation from window to view port, 3D geometry transformation and projection transformation. In a word, these algorithms help to convert the 3D model view into the computer screen and simulation the real visual habit of the human. Homogeneous coordinates is the most important concept in the graphic transformation. The principle of using this coordinates is to convert the n dimensional problem into the n+1 dimensional space. Main transformations include translation, zoom in/out, and rotation. In general, they can be represented by $R' = a_1R + a_2$, a_2 being in charge of zoom-in/out operations and a_2 in charge of translation, R being the original vector and R' the vector after transformation. In three-dimensional space, the vector is represented by a three-dimensional array. By adding a translation term, the array can be extended into a four-dimensional array. Therefore, we can extend the three-dimensional space into a four-dimensional space that combines the transformation of translation and rotation together. Another advantage of using homogenous coordinate is that the vanishing point can be identified in the virtual environment. For instance, $(ax, ay, az, a), a \neq 0$ with changing a represents a line in space. When a is approaching 0, the point can be used to substitute the vanishing point.

4.3.3 3D inquiry in OpenGL

Generally, it is very difficult to mark an object in a 3D environment, because that requires multiple transfer, rotate and projection operations. However, 3D space inquiry is one of the most important functions in GIS. OpenGL supports the mechanism of this function. The most important function in OpenGL is the selection mechanism, i.e. draw the current scene into the frame buffer, then draw the new scene in the select mode. The scene in the frame buffer is not changed when the select mode is being used. When the user quits the select mode, OpenGL can derive a list of intersection with the view volume. The selection mechanism is based on the select mode mechanism. Usually, it is activated by the input device and calls the function gluPickMatrix() i.e. use a special projection matrix to multiply the current matrix. By doing so, the matrix that we want to collected can be achieved. Through the select mode and the selection mechanism, the area that surrounds the motivation point can be selected.

The feedback mechanism is also a kind of input device motivation mode. In this kind of mode, the scene is not writing into the frame buffer, it is exported into the application program. The difference between the selection and feedback mechanism is that in select mode, an integer array is addressed, whereas in feedback mode a floating point array is addressed, which includes the information on the vertex, colour and other description parameters for the graphics.

4.4 Level-of-Detail technique for accelerating display features

4.4.1 LoD principle

The essential part of the Level of Detail (LoD) mechanism is to reduce the geometry elements of the objects that need to be rendered, while maintaining the basic attributes of the objects. The inspiration comes from common sense about human vision, i.e. when objects are far away they are not very clear to distinguish in much detail, but when they are very near, detailed information will appear. In computer simulation, the geometry factor is used to represent the objects. Three main components are used: points, lines and surfaces. We can combine these three elements to represent different natural objects. The triangularisation algorithm is optimized for 3D modelling so that lots of graphics acceleration cards are made based on that. The LoD technique is to reduce the number of triangles when simulating scenes, viz. to reduce the complexity of the scene while maintaining a high image quality (Röttger et al.,

1998). There are three main questions that need to be solved in the LoD algorithm:

- What is the appropriate data structure to store the magnanimity terrain data
- How to achieve a smooth transition between different levels of detail
- What node evaluation system can judge which nodes need to be subdivided

There are two main problems in quadtree based LoD algorithms for terrain simulation:

(1) T-Cracks, meaning the crack in the junction of the subdivided node if the difference of detail is larger than 1. The graph below shows the appearance of the problem: the black part represents the crack.

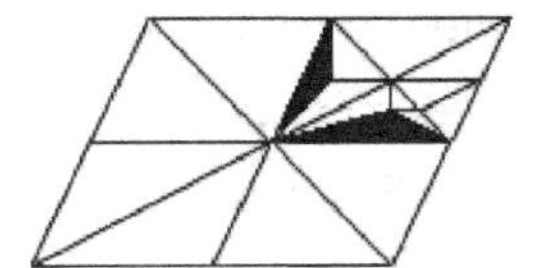

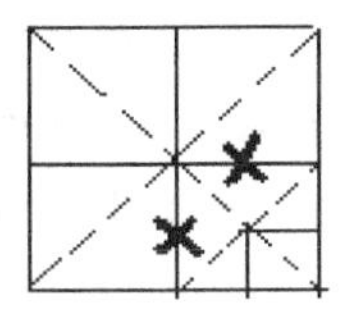

Fig 4.15 Crack problem (left) and Solution (right)

The solution of preventing this crack is to specify that the difference between neighbouring nodes cannot be larger than 1. To realize this, one should skip the centre vertex at the edge with the mark when rendering the grid. From the graph, we can see that the junction of the different detail levels can share one edge based on this criteria so that the crack will be discarded. The dashed line shows the triangle fans. (Fig 4.15)

(2) Popping, also described as the geomorphing that is caused by the change between different levels. Some parts of the terrain will appear suddenly, especially at the border of the terrain.

Although there are some fantastic algorithms to solve the geomorphing problem, the price to implement these algorithms is very large. For example, some people use interpolation to reduce geomorphing (Koller, Lindstrom et al. 1995). However, the cost of calculation time is too much and it is not good for real-time rendering systems. For our purpose, we just want to control our geomorphing in a mild way. Therefore, we adjust the threshold parameters C and C' of the node evaluation system (described below) to find the tolerant popping phenomena for terrain simulation.

4.4.2 Data storage method

We used the 2-dimensional matrix as the data storage structure, which can be easily coordinated with the quadtree data structure. Because the terrain node subdivision mechanism is based on the centre division regulation, the height map size should satisfy $(2^n+1)*(2^n+1)$ which guarantees the centre must exist. The graph below shows the raw data in the matrix and the representation of the real grid (Fig 4.16). The arrow sets the direction for subdivided the terrain node which is a top-down mechanism. On the right hand, the raw data is given. The entries 1 means this node needs to be subdivided, 0 means it needs not to be subdivided. The arrows indicate parent-child relations in the quadtree (from Röttger, 1998).

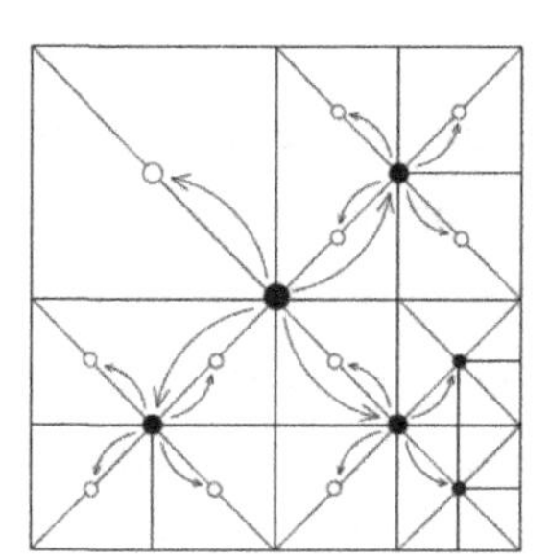

Fig 4.16 A sample triangulation of a 9*9 height field.

4.4.3 Node evaluated system

Nowadays, adaptive schemes are widely used for modelling complex objects. The idea of this dynamically changing scheme depends on attributes of different schemes to minimize the sampling of the geometry. Attributes can be classified as the criteria to decide whether or not to subdivide the scheme. In particular, the criteria can be concluded depending on the roughness of the terrain and the distance between the viewpoint and the terrain. The node evaluation system combines the two effective factors that construct the adaptive scheme for the terrain node. After the node evaluation system, we can know which node needed to be subdivided into the next level. (Röttger, et al., 1998) gave a very efficient method to build the node evaluation system, which was used here to build our own prototype system.

- Distance

The principle of the LoD algorithm is to have more detail when the viewpoint is near the terrain. Also, the size of the node needs to be taken into consideration. As show in fig. 4.17, l represents the distance between the viewpoint and the terrain node, d means the size of the terrain node.

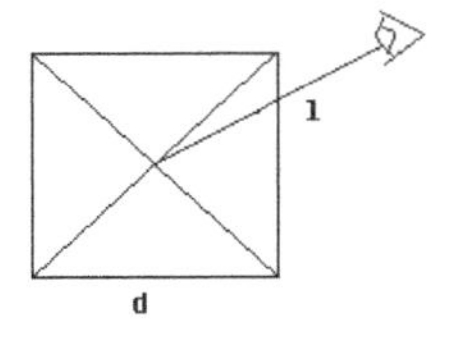

Fig 4.17 Distance Evaluation

$$\frac{l}{d} = C \tag{4.13}$$

C is a configurable quality parameter. When C increases, there is more detail need to be represented. Conversely, there will be less detail when C decreases.

- Roughness

We also should take the roughness of the terrain into account for the construction of the node evaluated system. For example, if the terrain is rugged, more detail will be presented. Inversely, a flat terrain has less detail. The roughness criterion is described by the following graph (Open, Dave et al. 2007). The graph below shows the 2D analogy to the 4-level terrain in (Fig 4.18).

L_0: {P_1, P_2}; L_1: {P_3}; L_2: {P_4, P_5}; L_3: {P_6, P_7, P_8, P_9}

The level error (d_i) is represented as the distance between the point in level L and the edge for level L-1. We assume that the roughness error for each level is constructed by this node's level error and its subdivided node's level error. In other words, when we calculate the roughness, we should take 2 levels into account.

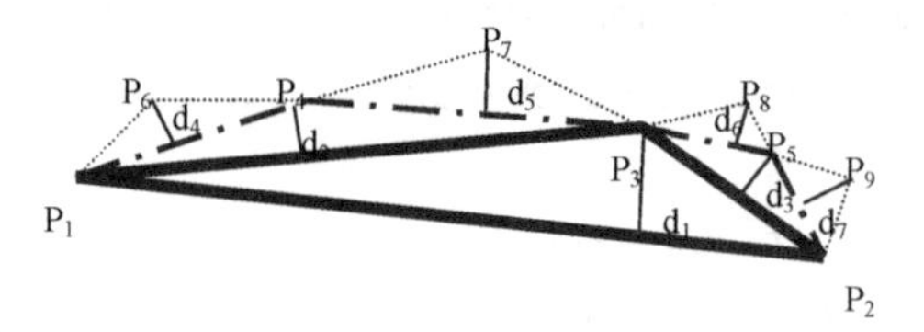

Fig 4.18 Roughness Evaluation

There are 9 key factors in this roughness evaluated system:

$(dh_1, dh_2, dh_3, dh_4, dh_5, Dh_6, Dh_7, Dh_8, Dh_9)$.

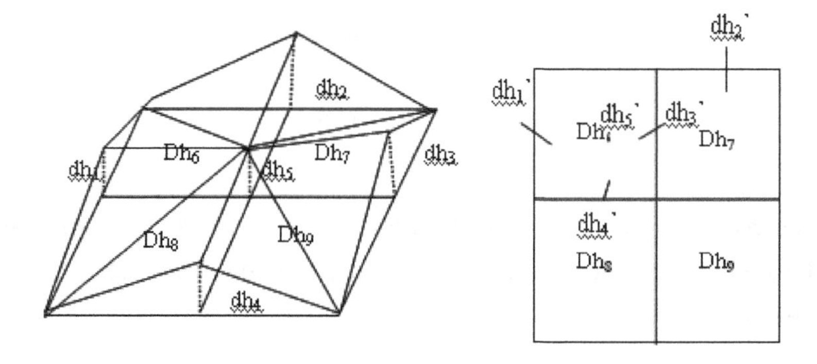

Fig 4.19 Iteration of the nodes' level error

dh_X means the level error for the higher level detail layer (more detail level).

Dh_X means the level error for the lower level detail layer (less detail level).

Since d means the size of the terrain node, the roughness evaluation equation is:

$$r = \max_{\substack{i=1,..4 \\ j=5,..9}}(dh_i, Dh_j)/d \quad (4.14)$$

$$1/r < C' \quad (4.15)$$

C' is another configurable quality parameter which is related to the roughness. When C' increases, more detail will be represented.

Now we can generate our equation for the node evaluation system which combines the factors of distance between the viewpoint and roughness of the terrain. When the conditions satisfy f<1, the terrain node needs to be subdivided

$$f = \frac{l}{d * r * C * C'} < 1 \qquad (4.16)$$

4.4.4 Basic Objective and Event Objective merged into Virtual Environment

The virtual environment, as the name mentions, should display the physical environment as realistically as possible. Basic objects and event objectives are two components for constructing the virtual environment (Gong J.H., 2002). In the previous sections, we discussed the optimal terrain simulation system that plays the role of the basic objects. Therefore, we want to extend a little bit to the event objects simulation, which can be considered as the surface information is merged with the terrain.

Water is the most important phenomena in the study environment. The construction of a water surface in a virtual environment is also a hot item in the field of visualisation. If we want to focus on constructing a virtual environment for flood control, where the information on the flood is the crucial element in simulation, the details or even the artistic representation of the water body like spray reflections and so on, will be not taken into account. However, the essential information about a flood event is the variation in water level. Therefore, there are two kinds of meshes in our system: the terrain mesh and the water layer mesh. The terrain mesh is as described before, representing the terrain contours. The water layer can be considered as a flat plane to merge with the terrain mesh. As soon as the water level is determined, the system will generate a flat plane into the 3D virtual environment to cover the terrain that is below the water level. We can give the flat plane the water texture or just the blue colour to reinforce the visual feeling in the virtual environment. In this simple way, we can view the flood situation very easy. In future work, we will link this water layer with the hydraulic model so that the water level will be the model result. In this way, the decision makers or the general users can have an impression of the future scenarios after the flood. Combined with different maps, for example the transport network map, the decision makers will know how to evacuate the residents in a short time.

4.5 Implement

4.5.1 System Design

The main parts of the visualisation system developed for the Yellow River Delta involve data input, rendering, 3D modelling, menu navigation and image morphing. From the data input and rendering parts, the users can construct a 3D version of the real landscape on the computer. The navigation part helps the user to determine his viewpoint and zoom in/out on the area of interest. The image morphing part can show the change in views in a dynamic way, for example exhibiting land use change or flood inundation extent, of course depending on the available data. If the texture file is detailed enough, users can see very detailed changes on the surface of the area. The structure of this part is described in the following paragraph. Changes in land use imply different conditions at the land surface. This implies that different 2D maps (texture files) have to be pasted onto the surface of the 3D landscape model. The flow chart of the system procedure is outlined in Fig 4.20

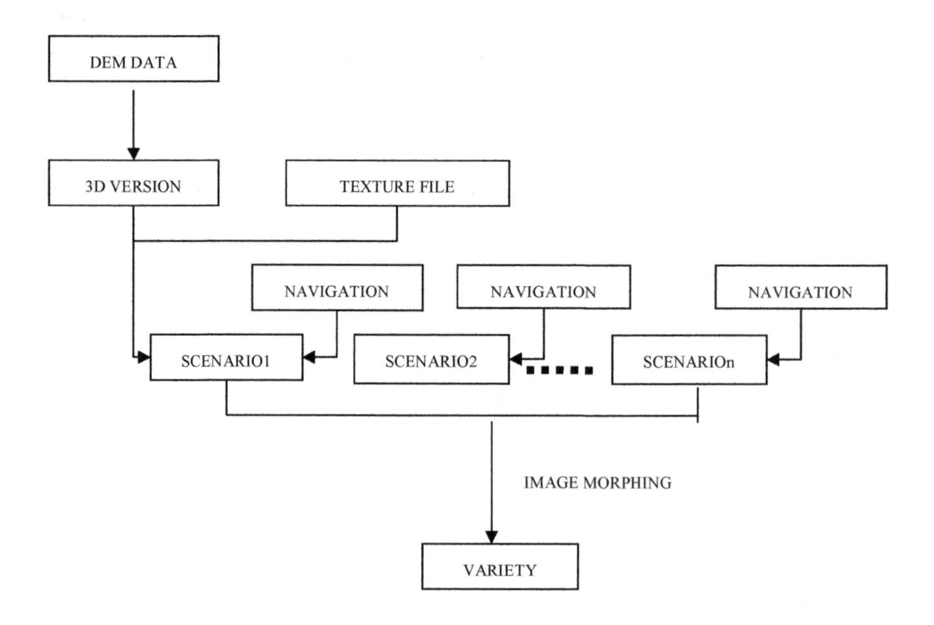

Fig 4.20 System Structure

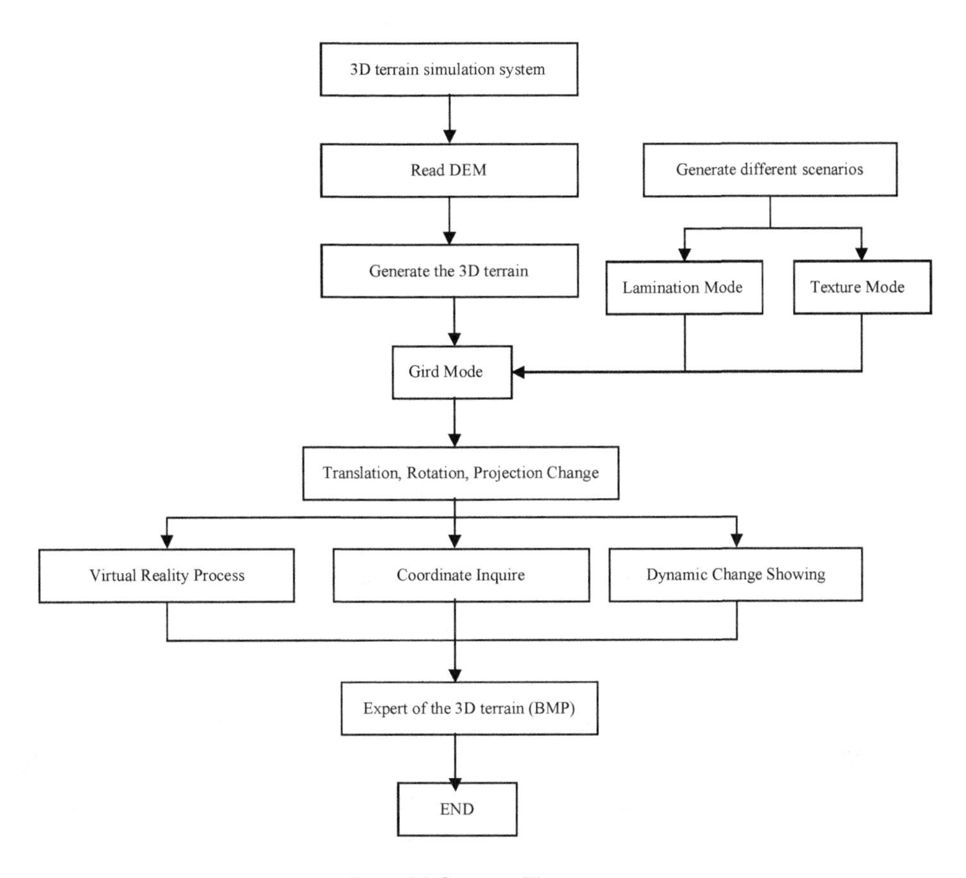

Fig 4.21 System Flow

4.5.2 Study Area

The Yellow River delta is characterized by the large volumes of sediment that cause the estuary to extend into the Bohai Sea, swing around and change course to create a fan-shaped region over the many centuries. It is located between north of Shandong Province and Bohai Sea. According to the difference in age and geographical position, the delta can be divided into the old delta and the new delta. In this thesis, the system will focus on the new delta, taking Yuwa as the apex from north of the river mouth, to south of Songcun Ronggou, that forms the fan-shaped region with an area of approximately 2400km2. Mainly, This area is largely man-made by changing the direction of the Yellow river in order to help the development of the river mouth economy, protect agriculture, and control floods.

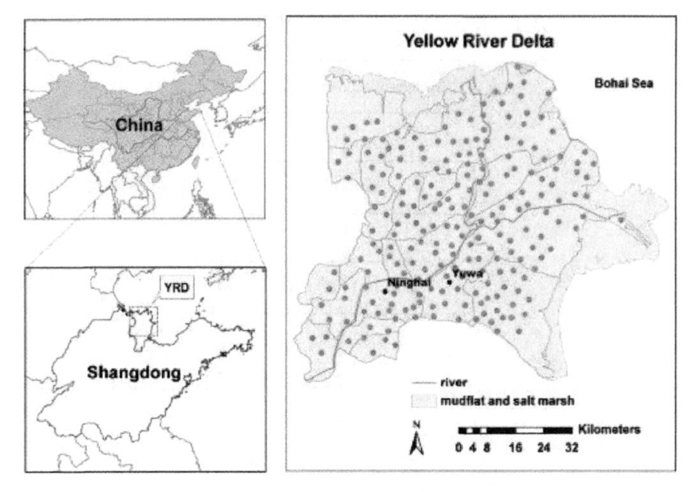

Fig 4.22 The Yellow River Delta

4.5.3 Effect of land use change and river discharge

The input data can be distinguished into two parts: DEM data and texture files. The DEM data was collected from the Yellow River Commission; the texture files were the particular land use maps as constructed by (Hui Ma, 2007) for the years 1986, 1996, 2001. The discharge in 1986 was slightly higher than 16 billion cubic meters, and the discharge for 1996 was about 19 billion cubic meters. The change in land use and extension of the mouth of the Yellow River delta are presented in Figure 4.23.

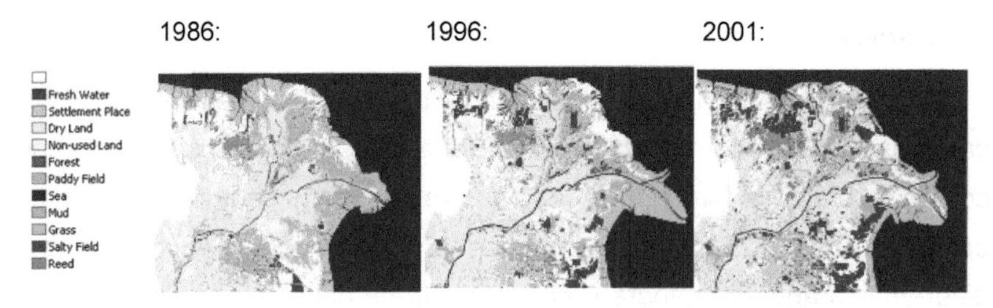

Fig 4.23 Land use change and Yellow River delta accretion

Future prognosis maps were constructed for two situations: Drought Period, and Regular Flood. According to the regulations formulated by the Yellow River Commission, the discharge for the Yellow River delta will be regulated to about 20 billion cubic meters per year (Wu Dong, 2007); if the discharge becomes less than 16 billion cubic meters, the year is identified as a drought. Based on this, predictions for future land use maps were generated for a drought and regular flood year as presented in Figures 4.24 and 4.25. It can be observed that the growing of forest in the drought scenario is not as good as in the flood scenario, which are indicated in the encircled areas.

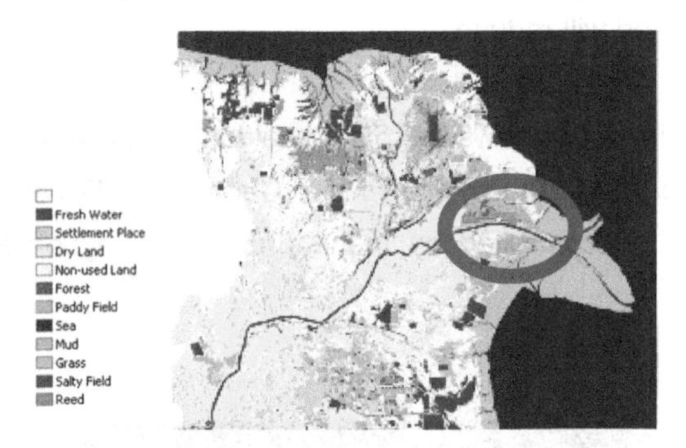

Fig 4.24 Future land use for drought scenario

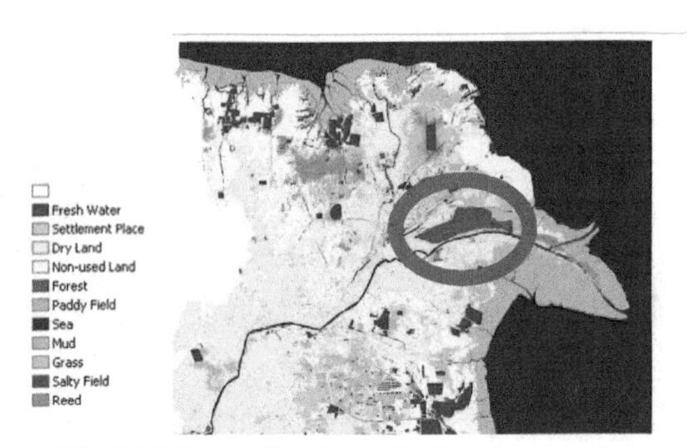

Fig 4.25 Future land use for flood scenario

4.5.4 Time display and morphing of maps

With the help of the Movie Maker part of the system, a short animation can be generated. The animation is constructed from the different scenarios' images which include the change of land use maps from 1986 to the future, is separated into a drought and regular flood period. From the animation, the users can see the changes for the different scenarios quite clearly. The animation can be pasted as texture files on the 3D land surface. The texture mapping technique in 3DS MAX can export the 3D landscape changes as an AVI file that can be actively studied by end-users. The users can make assumptions on the reasons why the change happened. In the case of the

Yellow River Delta, the fresh water availability can be directly related to the vegetation, especially to the plant growth near the riverbed. In case of regular floods, the forest and the reed areas near the river, represented by the dark green and the pink, are seen to expand. However, in a drought year there is not enough fresh water for the vegetation in the delta so that the area of the forest and the reed will reduce.

The image in Fig. 4.26 shows the 3D virtual landscape with the texture file of the aerial photograph of the Yellow River Delta. The 3D view angle and area of interest can be controlled by the end-user via keyboard and/or mouse actions.

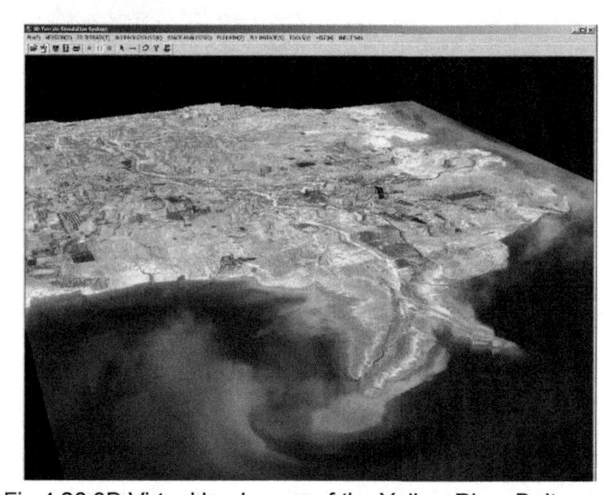

Fig 4.26 3D Virtual landscape of the Yellow River Delta

4.5.5 The effect of Level of Detail (LoD) on views and computational load

In order to explore the extent of flooding, a test case was carried out for different flood levels. The images in Fig 4.27 show the water levels in the virtual landscape in case of water depth increase by 1m, 3m, 8m resp. The LoD algorithm was used to reduce the resolution at some distance away, enhancing details as the fly-over path gets nearer, etc.

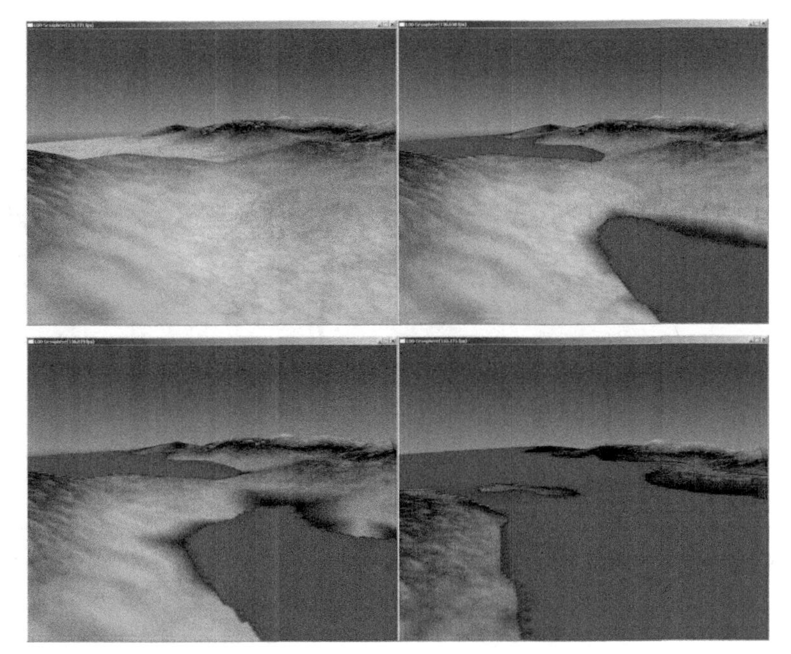

Fig 4.27 Virtual Environment with water surface in different water level (from left-up to right-down) h=0m; h=1m; h=3m; h=8m

A comparison of FPS(Frame Per Second) for different Level of Detail is shown in Fig 4.28:

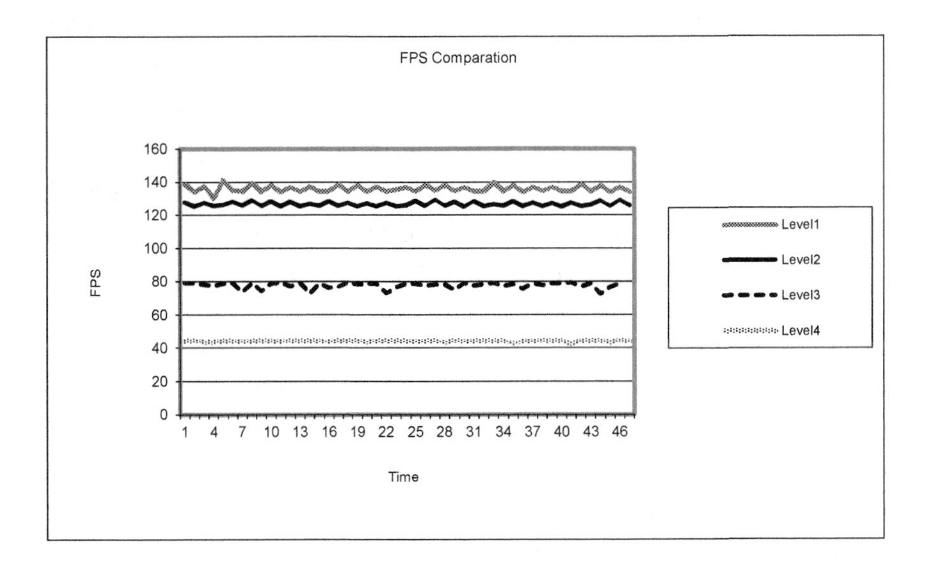

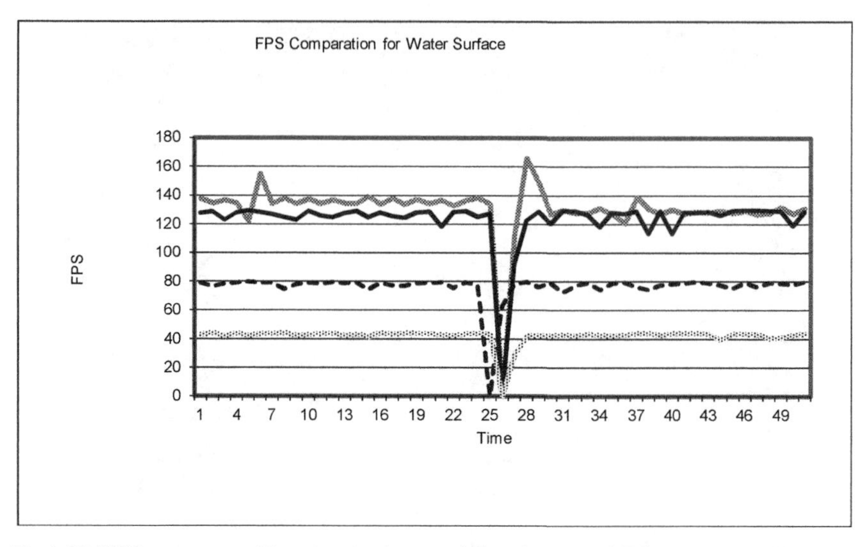

Fig 4.28 FPS compare without water layer adding (up) and FPS compare with water layer adding (down)

From Level 1 to Level 4, the level is increased which means that more detail can be shown. In this test, level 4 is the full resolution of the terrain data. One thing to note is that these levels refer to the start level: if the level is 1, the detail can be dynamically changed from 1 to 4 based on the node evaluation system; if the level is 2, the detail can be changed from 2 to 4; etc. From the test we can observe that in lower start level, the FPS is higher; however, the popping phenomena is also obvious. So we choose the middle start level to control the popping, meanwhile maintaining a reasonable FPS. The method for adding water levels does not have negative effect on the FPS values. The figure below shows the difference in FPS between no water surface and after adding water surface. The 0 FPS is the delimitation which means that users are inputting the water level into the virtual environment.

4.6 Summary

Using the virtual environment technique presented here, different scenarios can easily be rendered directly. With the dynamic views representation of the data, a considerable amount of data can be disseminated to end-users in an easy-to-understand way. Optimized methods can enhance the performance in order to provide very smooth images that are appealing to the user experience. This chapter demonstrates the capabilities of applying 3D graphics and virtual reality techniques in water related areas like land use change and flood / drought simulation. The study focuses on the 3D terrain generation technique based on DEM data which was implemented in a quadtree Level of Detail (LoD) model for accelerating the 3D scene generation for large scale terrain simulation. LoD facilitates having a quick response in the decision support process.

By data visualisation, the decision makers are likely to interpret model results more quickly and directly. This will help non-expert users to transform the information from numerical model simulation to a better understanding of the physical implications. By using the OpenGL library, the GIS inquiry can be performed in the 3D virtual terrain model. The basic information of the 3D space, for example the coordinates, the distance, the slope can be acquired form this application.

Interaction is the most important part in scientific visualisation. In the 3D terrain simulation system, the interaction allows users to zoom in/out, rotate, translate and change scales, which can assist users to view the virtual environment from different angles. Moreover, the interaction can be extended to more aspects. For instance, modifying parameters in the model and visualizing the simulation results can be very important to understand the sensitivity and meaning of this parameter.

5 Providing 3D Information in Flood Disaster Management

5.1 Introduction

Flood disaster management is an important topic especially in countries like the Netherlands and China who have been suffering from floods for a very long time. Disaster management is a complex process and faces a lot of challenges at the different stages: Prevention and mitigation, Preparedness, Response, Recovery and Post-Disaster (Pilon, 2006). During a recent training at the EU project training school named 3D Geo-information for Disaster Management (COST Training school report, 2009) there was a group assignment about designing a generic flood disaster management system where it became apparent that, because of the different background and knowledge of those involved, each had its own point of view about such system design. Participants came from different disciplines related to disaster management: cartography, ontology-based design, hydrodynamic modelling, spatial decision-making, human-computer interaction, 3D spatial data modeller and hydroinformatics research. This combination of backgrounds covered almost all types of actors in the risk management area (not on emergency response, unfortunately).

However, there were difficulties identifying the generic process that could take into account all types of flood disasters in all countries. During the discussion, some participants had strong doubts about the effectiveness of flood early warning systems. They preferred more attention to actions during emergency response. This discussion seems to repeat the debate about the effectiveness of flood early warning (Drabek, 2000; Handmer, 2002; Parker, 2003) and on the role of computer models and information systems (Sene, 2008). Their research suggested that the problem of providing flood warning to the public was mixed. On the one hand, it is important to develop advanced flood forecasting models to improve the lead time and accuracy of the warnings; on the other hand, communicating the warning information, arranging community participation and collaborating between different agencies, is becoming more and more important. From the author's point of view, a flood disaster management system should combine physical modelling techniques, social issues and economic aspects together. Information sharing is at the heart of such system, and advanced visualisation can help achieving this.

Nowadays, flood disaster management systems actually can be considered as digital information systems. Most of these systems are using 2D information representation. However, 3D information has the advantage of improving the communication process between people with different knowledge background. In this chapter, the assumption was made to design a flood disaster management system as one software system. Therefore, we utilize software engineering methods to describe the whole system. First, the principles of disaster management are introduced and the requirements for a flood disaster management system are specified. Then, the focus is on the role of 3D information and visualisation during the communication process. After that, a generic model for flood disaster management is introduced based on UML modelling tools. In this model the focus is on the possibility of utilizing 3D visualisation data options. At last, a discussion is presented on how crucial information can be made available in flood early warning systems that can guide the decision making process.

5.2 Requirements in Flood Disaster Management System

5.2.1 General Description

Supportive technologies are very important for developing a flood disaster management system. The principle is that within effective tools, the information can be easily updated and the system should be flexible enough to develop scenarios and provide visual and quantitative information regarding the forecast event (Pilon, 2006). Geographic Information Systems (GIS), Mapping techniques and Visualisation all contribute to this principle. Among them, visualisation is encapsulating the other two technologies and acts as the portal for the (disaster) management system, which provides direct interaction with the end-users.

With the development of computer graphics, real-time rendering of 3D scenes has become possible. The concept of virtual environments is more commonly accepted because of the advantages in representing information directly and supporting interaction between the user and the system. However, different areas, e.g. computer graphics, GIS, Architecture, Engineering and Construction(AEC) and computer gaming, have different requirements towards virtual environment technologies (Kolbe, 2006). Generally, there are three main requirements in disaster management system design towards the particular virtual environment: Interoperability, Communication and Dissemination (FEMA, 2012). The essence of interoperability is how to share and coordinate data from different sources. One example of Interoperability is the development by the ministry of Infrastructure and Environment (I&E) in the Netherlands of a central OGC (Open Geospatial Consortium)-based services architecture infrastructure that allows data sharing amongst all users in a distributed system (Grothe, et al., 2005). Communication facilities are built on top of the data sharing environment to assist different planners and actors understand flood risk which means generate the same Cognitive Knowledge about what happened in the past and what will be the effect of measures in the future. Since often communication problems exist in the interaction between humans and systems, the UML case diagram can be useful to illustrate the communication requirements between the system and its external environment. Norliza (Norliza, 2009) used UML and ontology analysis for designing a web-based support system for flood response operations in Malaysia. He focused on the emergency response phase of disaster management systems. However, the relationship with the other modules, like flood forecasting and flood warning, was not mentioned. Dissemination in a decision-making process is about assisting decision makers to be aware of the situation and have all relevant information to choose the best decision in an emergency condition (Klimenko, 2008). Most of the information that is usually presented in front of decision makers is geo-information. Many countries have started developing spatial information infrastructures for emergency response and early warning. However, most of these developments are still limited to 2D-maps (Zlatanova, 2008). In the next two sections the system requirements are derived for two cases, as described hereafter.

5.2.2 Case of Netherlands

In the Netherlands, preventing flood disasters has a very high priority. Due to the low-lying land, the high water threats are taken into account from the very early stages onwards. Recently, there were three flood disasters that affected the country very much: the 1953 coastal flood, and the 1993 / 1995 river floods. The 1953 flood was caused by a severe storm surge along the coast. The other two were related to heavy rainfall causing the river levels to rise. After the 1953 flood, the famous 'Deltaplan' was established for seawater defence by building dams, sluices, and storm surge barriers. The river floods in 1993 and 1995 gave a clear view of the importance of flood management. During the period of high water two items were of considerable importance: early-warning and preparedness-for-evacuation (Bezuyen, 1998).

In case of a flood disaster, the municipalities are responsible for disaster management in the Netherlands. There are two levels of emergency management in that case: regional and local level. The regional level is activated when the municipal boundaries are transcended. Mayors are the main decision makers at the local level, whereas mayors cooperate with the provincial governor and minister at the regional level. Fire brigade units play an important role in both local and regional emergency planning. In the Netherlands, the fire brigade not only deals with fires, but is also responsible for the on-site coordination. In case of floods, several other organizations also take actions when the operational organisations need support. For example, the Directorate-General of the Ministry, the Dutch National Reserve, the Royal Dutch Water Life Saving Association (KNDRD), the Royal Netherlands Sea Rescue Institution (KNRM) and the Search and Rescue units (SAR) (Arta Dilo, 2008). The Ministry of Infrastructure is responsible for communicating the water level predictions along the Dutch rivers and for supervising possible evacuation plans.

Depending on the warnings issued, the different management units can be activated. Then, traffic management, medical assistance and other support can be arranged. The disaster plan is updated every four hours to describe the most important threats and define the corresponding emergency activities. The flood warning system is the trigger to activate different actions and secure proper response. Support systems for spatially distributed decision making have to take into account for four main items: identifying the locations of the critical areas, exploring countermeasures, deciding on specific actions, organising implementation of these actions. Visualising the spatial database is very important in the communication between different organizations and actors.

5.2.3 Case of China

China has a huge land area of some 9.6 million square kilometres and approximately 1.3 billion citizens. Due to its specific geographic location, the effect of monsoons and tropical cyclone conditions, China has been suffering from flood disaster for a long time already. After the establishment of People's Republic of China in 1949, flood disaster management was also changed because of the new political structure. River basin authorities were established for all major river systems in the country. At present are 7 river basin commissions bureaus directly under the Ministry of Water Resources. Among them, the Yangtze River Commission and the Yellow River Conservancy Commission are at the vice-ministerial level while others are the departmental level. The national flood disaster management system is a huge system due to the scale of river basins. Besides the 7 river basin bureaus, there are 16 structural departments, 31 provinces, 31 important cities and 100 large and medium reservoirs. When a flood disaster is about to happen, there are several hierarchies in the system: ministry level; river basin level; province level, autonomous region level, municipality level.

For different river basins and municipalities, there are different flood control plans that have been built with the help of simulation and optimisation models. Normally, all involved follow these plans and participate actively in the flood control operations. However, when the situation becomes extreme, the lower level should report and discuss the situation with the higher level. Especially when the situation reaches the flood control departments, the decisions must be communicated at the ministry level. Because there are lot of residents living in flood prone areas in China, decisions must be taken with care. For example, when there was a huge potential flood in 1998 along the Yangtze River, the flood control plan prescribed the flood division in the Jing river division area should be activated. However, the ministry level decided to suspend the plan meanwhile evacuating the residents as soon as possible. Fortunately, the water level dropped afterwards.

There are 3 types of actors in the emergency response phase in china: the military, mechanized flood fighting teams and temporary teams composed of residents. The second actors are of a new type since the reform of the water management structure. These teams have advanced mechanization which are specific for flood control activities and the team members include flood control experts. Flood forecasting is very important in the flood disaster management system since it affects the flood warning directly. China developed its own flood simulation models with empirical formulae for flood forecasting. Spatial information and database are taken into consideration more recently with the development of digital techniques, e.g. GIS, RS and GPS. Also here, communication between different actors and decision makers is considered very important.

5.3 Role of 3D Information

As can be concluded from the previous examples, flood disaster management is a huge topic that includes many different aspects. As a general description of the process, quite often five phases are distinguished: Prevention and mitigation, Preparedness, Response, Recovery and Post-Disaster (Greene, 2002). Normally, these five phases can be grouped into two categories: (i) risk management and (ii) emergency response. In this research, we identify a third category, i.e. (iii) a flood early warning system for flood disaster management. From the time requirement point of view, the warning system plays a role between the preparatory planning phase and emergency response. It is the trigger to emergency response activities. GIS is increasingly used in both groups because of its effective visualisation capabilities of disaster situations. By placing the actual physical geography of a disaster event on a computer monitor and then combine with other relevant features, it becomes more 'natural' to make decisions based on GIS information (Snoeren, 2007). Spatial Data Infrastructure (SDI) is an important component that facilitates and coordinates the exchange and sharing static and dynamic spatial data. The actors in the two groups of disaster management identified above, are people who have the most contact with the spatial data so that the requirements coming from them can determine the design of the SDI. Traditionally, the spatial data is represented in 2D formats while the user interface of a GIS system is used as the main UI throughout the whole management process. With the development of computer graphics and information techniques, 3D models and 3D visualisation are available to serve a more real representation of spatial data. However, what is the role of 3D information and how to merge this information format into the existing data infrastructure? Therefore, we have to specify the main actors and activities among the three groups so that we can determine the real demand for 3D information.

5.3.1 Flood Risk Management

Risk is the estimation of the impact from a disaster. There are three main components in risk definition, i.e. hazard, vulnerability and resilience. Hazard is the disaster that can cause the impact - flood is such kind of hazard. Vulnerability is the susceptibility to suffer damage from the particular disaster hazard. This part is dynamic and related to the human activities to protect themselves and adjust themselves for a certain hazard. High vulnerability refers to high impact of a hazard. Resilience is a relatively new concept that refers to the physical and social processes that may affect hazards and vulnerability, e.g. new ways of urban development. Risk maps are a way to visualize spatial information with a certain risk index. It is also a mapping technique to generate a graphical image with the level of risk. The producers of these kinds of maps are geographic information experts, simulation modellers and other relevant disciplinary staff. Flood hazard maps define the inundation process of the flood and flood damage maps provide the information on the potential damage with a certain probability of occurrence. Fig(5.1)

Fig 5.1 Example of flood hazard map (from GeoTerrain Tool)

Flood risk maps express the number of casualties under a certain flood event and the corresponding economic loss. This map is the guideline for subsequent actions. In the planning phase, the flood risk map is generated from historical thematic event maps. In the flood early warning system, the map is built based on real time data. Risk management has a close relation with urban planning and regional development that consider risk factors and vulnerable objects very much. The main activities in flood risk management are identification of potential risks, evaluation and assessment of flood risks, decisions on the implementation of risk reduction measures, monitoring and maintenance. Actors are civil engineers, transportation engineers, urban planners, housing organizations and so on. For flood risk mapping, 2D format can satisfy the requirements from civil engineers and transportation engineers because they are familiar with this type of representation. However, it is more difficult for urban planners and housing organizations to get all information from a 2D risk map. They would prefer a system that has tools for walk-through and fly-over and allow different views in which 2D/3D graphics and images can be observed simultaneously (Zlatanova, 2008). However, time constraints in risk assessment are not very crucial, since decisions can be made in a long-term process, involving different experts evaluating different alternatives and pursuing acceptance of all stakeholders. Many advanced models can be implemented in this phase. Therefore, detailed 3D models can be built in this period with 3D visualisation techniques to realize the interaction between human actors and the flood management system. Then, these 3D models can be put into a database as a tool for emergency response.

5.3.2 Emergency Response

Emergency response (ER) is the most crucial stage in the disaster management and puts high constraints on interaction time. As the two cases mentioned in the previous section, the actors who work onsite on the emergency response mainly are fire brigade, medical assistance, police, military and so on. Their tasks are assisting evacuation, saving people and protecting the flood defence structures. The group of people working in the control room are mainly experts and key decision makers. Their main jobs are containment and control of the flood and its effects. The relationship between these two kinds of actors is like 'eyes and hands'. The main decision makers decide whether or not evacuate, the actors onsite are in charge of implementing the commands.

In general, there are three kinds of coordination in ER activities: strategic coordination, tactical coordination and operational coordination. This three-tier hierarchy is suitable for every specific flood disaster management case. The strategic coordination mainly refers to the government level decisions; the tactical coordination is the regional command maker for onsite actors; and the operational coordination is with the local actors. The representation of information throughout this hierarchy should be different because the actors have the different concerns at their corresponding level. For example, the control room needs to survey the overall situation whereas the fire brigade is concerned more with particular areas or buildings. Although, the requirements for the different groups are different, the basic question they want to solve is how high the water level will be and when the inundation will happen. A GIS system with a flood simulation model is good at representing and predicting the flood in 2D format for displaying the water level variations. However, it is not enough for incidents situations because of the requirements of more detailed description, e.g. on what is impact of this water level and what is the inside the buildings and so on. Here 3D visualisation has the advantage of detailed and real representation of the scene so that the users can view the (potential) impact of flood disasters 'with their own eyes'. This is important for operational coordination because the incident they will handle is in the eye-sight range. More knowledge about a site means more effectively completing their tasks.

5.3.3 Demand for 3D Information

In disaster management, the demand for 3D visualisation and models for emergency response is growing (Zlatanova, 2008). Research reveals that the results are very much influenced by the familiarity of the users with the manner of visualisation. Most of those involved in the emergency response phase are not familiar with GIS and require another direct representation method to depict the situation as fast as possible. However, although 3D visualisation within Google Earth is considered important by the majority, the most web services for 3D virtual city modelling only support graphics or geometric models, neglecting the semantic and topological aspects of buildings as the terrain is being modelled (Kolbe, 2006). It is not enough to only show fancy 3D scenes to the system users, but also to present relevant information about the models which can support queries. The role of 3D information is not isolated as a representation method but integrated with other formats or data types in order to facilitate a flexible spatial knowledge retrieval process. We cannot say what information must be in 3D or not, which depends on the users preference and knowledge level. In system design, we should loosely couple the 3D data type with the data in order to allow flexibility in the user's choice. Therefore, if 2D images are enough for disseminating situations, we keep those parts in 2D. Possible 3D information parts should be identified and different models are to support the information representation. There are some classic 3D models for arranging the 3D data: simple box models for buildings, indoor models, underground models and terrain surface models. We should identify different models for different purposes in a flood disaster management system. Meanwhile, we can use the attribute data as the semantic information representation part that can link to the 3D objects. Because flood early warning is the trigger to emergency response, we also consider 3D information as the output of this module in order to send the information to the emergency response actors.

5.4 Object-Oriented Design

5.4.1 Concept of Object-Orient

The main differences between Object-Oriented (OO) and traditional procedural declaration computer languages are the concepts of Inheritance and Polymorphism. Inheritance represents the relationships between 'Family Classes', which means the 'Child Class' that can inherit attributes and methods from the 'Parent Class'. Upon implementation, these characteristics facilitate the resource saving and explicit representation of nested relationships. Modellers firstly define the common attributes and methods about the class, and then abstract them into the 'Parent Class'. Therefore, those attributes and methods only need to be written once in the program. By defining the 'Child Class' inheriting from the 'Parent Class', those attributes and methods can be used by the objects of the 'Child Class' freely.

Polymorphism is the ability of objects to respond to the same message with the appropriate method based on their own class definition (Poo, 2007). Different Classes can share the same method names, but point to different functions during execution. The message sender knows what to do but does not know how to do it. Polymorphism encourages the modeller to focus on specifying the properties and the functions of the Classes without apprehension over how to realize various events. Fig 5.2 shows the conceptual structure and difference between procedural declaration (PD) and object orientation (OO). PD enumerates different kinds of plants and isolates those plants in independent structures. In contrast, OO emphasizes the abstraction of common characters and the relationship between groups. Those common characters have been encapsulated into the 'Parent Class' that can be inherited by the 'Child Classes'. Therefore, declaration duplication is much reduced in OOP. It is a flexible structure in which new groups can be added to the system in OO. Another advantage of OO is simulating discrete groups of objects that have the same characteristics and spatial patterns. OO is good at describing complex systems which have nested structures.

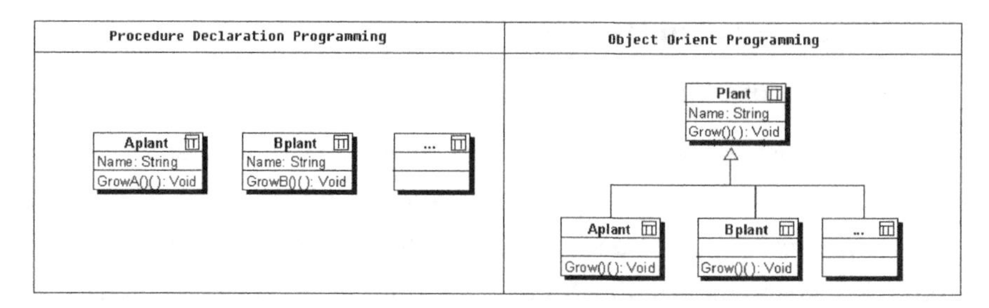

Fig 5.2 Concept Structure. Left: procedure declaration programming; Right: object oriented programming

5.4.2 Concept of UML

The UML model is one of the standards to depict the system architecture in the software engineering area. The advantage of this modelling tool is that it uses the explicit visualisation method to represent the system. It uses the object-oriented architecture to construct a conceptual model of the system. Structural diagrams and Behavioural diagrams are two main components in UML that can correspond to static and dynamic processes within the system life span. Structural diagrams emphasize the objects that must be represented in the system. The Class diagram is such structural diagram and widely used to display system components and their relationships. Behavioural diagrams are used to describe the functionalities of the system which means what the system must do. The Case diagram is one such diagram to show the actors in the system and the dependencies among the cases.

5.4.3 Conceptual Design for Generic Flood Disaster Management System

For two main actors in a system, i.e. observer and actor, the use-case diagram is shown below (Fig 5.3). Two members can inherit actors: the Disaster Manager in the control room and the Crisis Coordinator onsite. The disaster manager can be considered as the main decision maker who takes the whole view of the situation. The Crisis Coordinator represents the group of people who execute the commands from the control room, e.g. fire brigade, police, ambulance, and municipal structures. Table 5.1 shows the definition of each use-case.

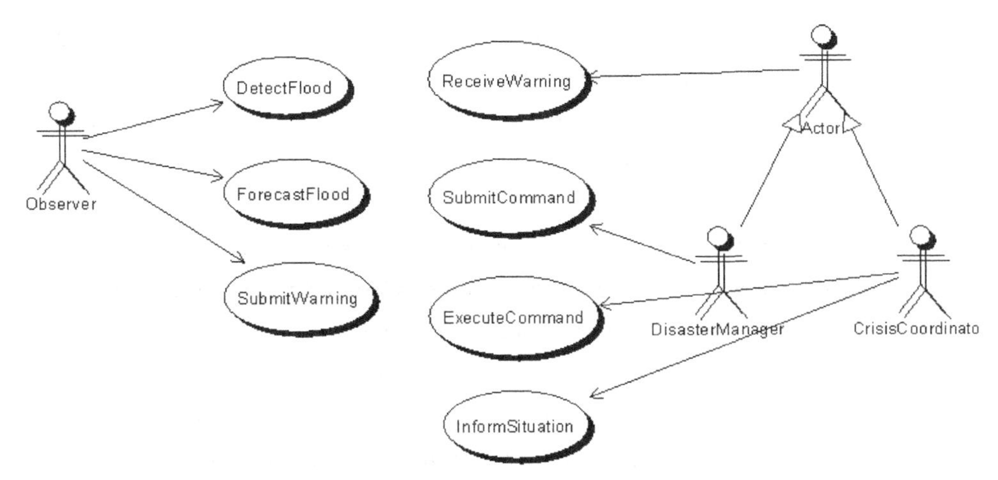

Fig 5.3 Use Case Diagram for the concept design

Use Case	Definition
DetectFlood	Using sensor system to detect meteorological data
ForecastFlood	Using modeling technique to process the data for forecasting future situation
SubmitWarning	When future scenario is expected to be high risk, flood warning message will be submit to actors
ReceiveWarning	When the flood warning message is received, multiple information will be processed in order to generate the command
SubmitCommand	Main decision maker masters the complete information about the situation and in charge of submitting general commands
ExecuteCommand	Executing the general commands
InformSituation	Informing the in-situ situation

Table 5.1 Definition of each use case

We divided the flood disaster management process into three parts: risk management, flood early warning and emergency response (Fig 5.4). The main difference between these tasks is that time is the most critical in the emergency response phase. The decision of risk management is based on long-term procedures for planning, which means the flood risk is only one part of the whole risk management process. Flood threats are what early warning systems need to address. On the other hand, incidents caused by floods are what emergency response needs to take into account. Since weather forecasting techniques become more and more mature, the accuracy of flood forecasting has improved dramatically compared to other early warning systems, e.g. earthquakes.

Generally, there are three components to construct a flood early warning system: hydrological detection, flood forecasting and flood warning. Detection includes the sensor systems and other method to monitor the weather conditions in order to support the flood forecasting system. Flood forecasting includes models to generate future scenarios in terms of inundation maps, velocity distributions and water level variation. FEWS (Deltares) is such type of system that supports flood forecasting in many parts of the world. The reason why we separate flood warning into an independent module is to reduce the complexity of the objects in flood warning dissemination. These can be the decision-makers, regional and local water managers, the public, press etc. However, the actual flood-warning message is the trigger of the emergency response process.

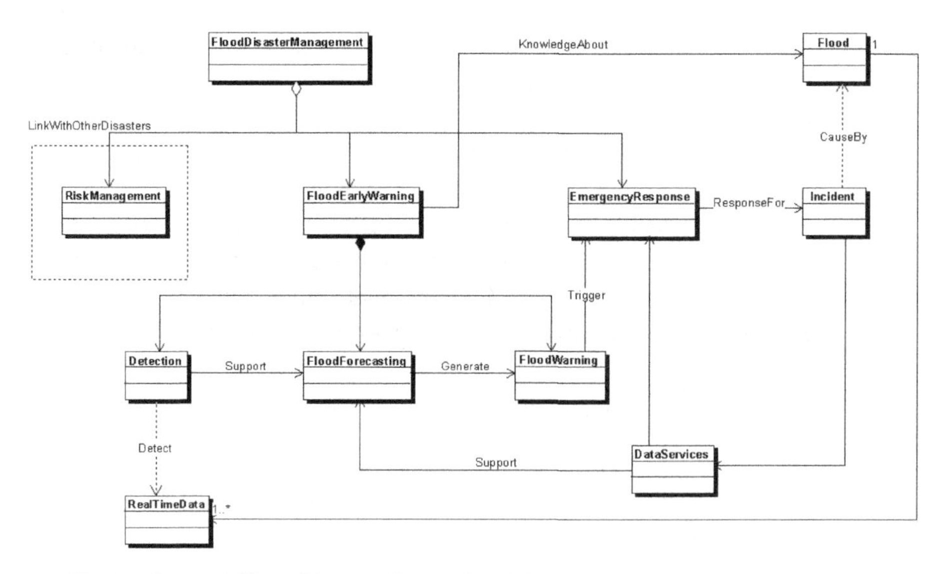

Fig 5.4 General Class Diagram for the flood disaster management system

The real time warning on a possible flood event comes from monitoring systems (Fig 5.5). There are two components in this data class: meteorological and hydrological conditions. They can be detected and filtered by the detection system and be prepared to support the flood forecasting system. Actually, the data corresponds to the requirements from the flood forecasting models. Most of the data are represented in 2D maps, ready to be exported. Most users of this kind of data format are familiar with GIS so that the 2D data type suits them very well.

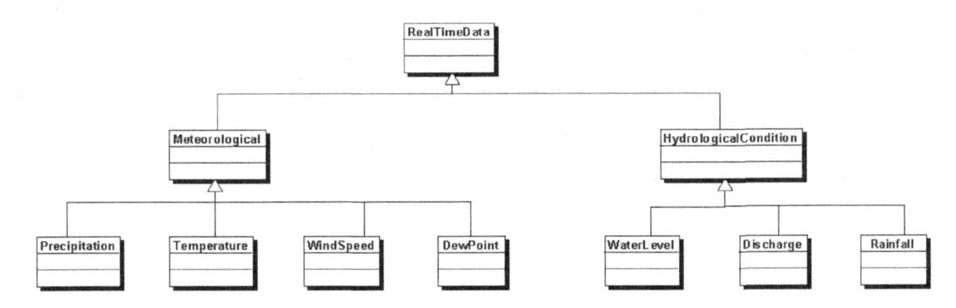

Fig 5.5 Detail Data Model of RealTimeData Module

The data service module can be considered as a database system that includes both static and dynamic information (Fig 5.5). It is linked to the incident module, which means that the data are both operational (dynamic) and existing (static). The dynamic information is mainly sent back by the emergency response actors on site, e.g. casualties, accessibility, flood info surrounding the incident, and so on. The static information contains reference data, demographic data, infrastructure data and so forth. In this section, we describe the data service module by the different data types and the data utilities. There are four data types: Map, Attribution, Image and 3D (Table 5.2).

DataType	Definition
Maps	Contains the location and shape of geographic features. Raster and Vector data models are used to describe this kind of spatial data
Attribute	Refer to the properties of spatial entities, normally are tabular data
Images	Raster data includes satellite image and aerial photos, for supporting more information to the maps
3D	Detailed data model to describe the objects. Surface, SimpleBox, Indoor and Underground models are used to execute this implement

Table 5.2 Definition of each Data Type

The data are mostly collected from five components in the flood disaster management system: Geographic, Hydrographic, Transportation, Construction and Socio-Economic aspect. The diagram (Fig 5.6) below shows how the different data corresponds to the different utilities and the structures of the data service module. The data type parts we set to the interface stereotype and the five data utilities are the instances of the data service module. By doing so, the data in the utilities can choose the appropriate data type for representing the information.

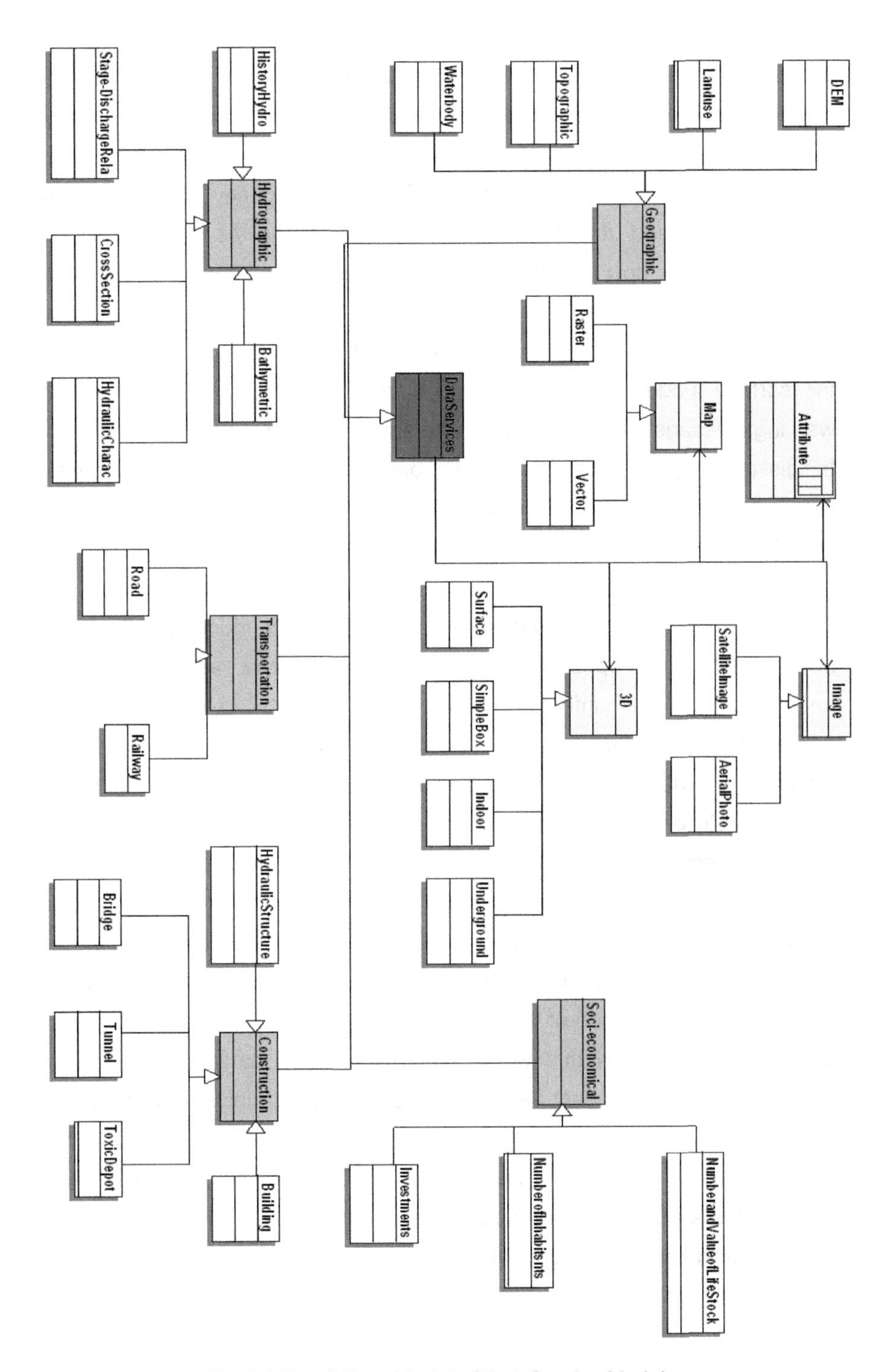

Fig 5.6 Detail Data Model of DataService Module

As mentioned before, the advantage of 3D information is particularly relevant for the communication phase and includes the flood warning and emergency response modules. Flood warning generation involves information conversion from the model to the operational procedure. It is this information that triggers the emergency response phase so that the emergency response actors should understand the content of this information as well. The main information is on 'how high the water level will be where?' What is the impact area? When it will come? How long will it stay? Therefore, the representation of the flood warning module can be rendered as a 3D scene which includes the surface model of DEM data and water body (Fig 5.7). After the flood warning message has been issued, the emergency response phase will start. In this module, the participants are mostly non-GIS experts. An integrated virtual environment such as Google Earth is suitable for them to illustrate situations and discuss alternatives.

On a large scale, the combination between the surface model and the simple box model can support the emergency response actors in quickly understanding the information around incidents. With progressing urbanization, more and more underground infrastructure has been built requiring knowledge on underground features. Therefore, an underground model should be included in the 3D data type of the data service module to grasp the complete situation. The underground model can be merged at the same level of detail as the simple box model. The interior of a large building is very important for navigating rescue forces to the place of the incident and supporting the evacuation plan to the people. Indoor models can support a more detailed description of the buildings and maximize the level of detail in a virtual environment. In this virtual environment, the data utilities focus on topography, flood impact area, transportation, structural aspects and demographic data. Meanwhile, the semantic information about the objects is also very important to make decisions. All 3D data types can participate in this module: the surface model for terrain simulation, the simple box model for outside buildings, the indoor model layout of room situation, and the underground model for the built environment (Fig 5.7).

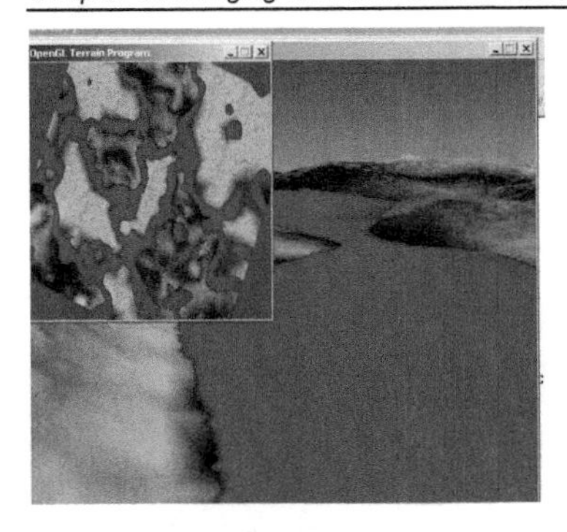

Fig 5.7 Different 3D model for spatial information representation:Up- Left: surface model(Xuan Zhu 2009); Up-Right: simplebox model(Zlatanova, S); Down-Left: underground model(Zlatanova, S) Down-Right: indoor model (Thomas Kolbe)

5.5 Summary

Flood early warning plays an important role in any flood disaster management system. Using the UML concept for model design makes it possible to understand how the data and information should be processed in order to generate a clear view of how the flood disaster management system is working. Class diagrams can be used to describe the data types and data utilities in the data service module so that each data entry can be matched easily with the appropriate data type. Often 3D data is used in sending flood warning messages and in the emergency response phase. However, it is not always necessary to use 3D spatial information throughout the whole system, but this very much depends on the requirements from the users whether their disaster management task is rational and economical. In the 3D spatial modelling part, four types of models are used (surface, simple box, indoor and underground) to represent the spatial information. Among them, the surface and simple box model are already developed maturely, whereas, the indoor and underground model is still under development. The semantic information about 3D objects also needs to be contained which can be implemented in the design by combining 3D and attribution data types. In future, we should look for more flexible methods to merge semantic information with the 3D data model.

6 Web-based Visualisation and Interaction in the Early Planning Phase

6.1 Introduction

Coastal areas have long been attracting human activities. In particular in countries like the Netherlands, with a large part below sea level, coastal flooding is one of the main threats requiring continuous attention on coastal management strategies. The Holland Coast is important for the Dutch economy because of its wide variety of economical, ecological and recreational functions. Those functions put a high pressure on spatial planning and management of this area. To prevent coastal flooding, the current strategy is to maintain the coastline at its 1990 position by annually supplying sand for beach nourishment from the deeper North Sea. Each decision on changing the shape of the coast can affect a large number of stakeholders at different levels, including central government, provinces, municipalities and water boards, as well as the industrial sector and other stakeholders including non-governmental organizations.

Traditionally, specialists in coastal morphology and hydrodynamics play an important role in advising on coast-related decision making; they may, for instance, assist decision makers in designing effective coastal protection measures or in assessing the impacts of alternative measures (Koningsveld 2003). Such decision making process can be represented by an inverted triangle where the wide edge represents the broad participation of the public and the vertices point towards submitting alternatives to specialists for their assessment. One disadvantage of this process is that the alternatives reaching the specialists may not be based on adequate domain knowledge and information on specific coastal processes. Communication and interaction between specialists and end-users become important factors for connecting the specialists earlier in the decision making process. The advantage of consulting expert knowledge as early as possible is that it can make individual stakeholders aware of the implications of their preferred solution on other domains.

Coastal management information is frequently communicated to stakeholders and the public through complicated management documents and engineering plans (Jude, 2007). Visualisation can provide a more suitable format to support the communication and dissemination process towards the public. In chapter 3, an online map-based user interface was designed for accessing alternatives. Web 2.0 techniques were used and proved to be efficient for defining future scenarios. A general framework was presented for visualizing model results and interacting with advanced simulation models. Is it possible to transfer this methodology to spatial planning? That is the topic of research in this case study.

Goodchild (2007) suggested that the digital earth can be considered an 'experimental environment' for spatial planners. The recent contributions by digital earth users support this vision. The Internet has become an efficient method to communicate between the general public, spatial planners and decision makers. It plays the role of a platform for comprehensive information representation and discussion of alternatives (Rinner, Keßler et al., 2008). However, how to use these techniques in coastal zone management with public participation is new, largely unknown, and seldom reported.

This chapter introduces a web-based tool for integrating data, models and tools at the early planning phase. The main objective of the tool is to stimulate interaction in public participatory decision-making. Using appropriate visualisation and interaction techniques, expert domain knowledge could be transferred to end-users. From a system design point of view, the loose coupling and generic format definition facilitate expansion of the user interface module.

6.2 Collaborative Geographic Applications

The term 'collaborative geographic applications' appeared in the early 1990s when various researchers started to develop group-decision-making systems. It was shown that basic GIS techniques could be used in innovative ways for comparing the outcomes of multiple stakeholder groups (Faber, 1994; Faber,

1995). A collaborative environment was created for experts and stakeholder groups. The process involves experts explaining their scientific knowledge and stakeholder groups stating their experiences and preferences (Jankowski and Stasik 1997). This is strongly related to the concept of public participation processes. According to the definition used by the international association of public participation (2006), five levels can be distinguished: inform, consult, involve, collaborate and empower. The lower levels of participation such as inform and consult, require information representation tools, e.g. maps, photos and hypertext style websites. The main purpose at those participation levels is to assist participants becoming aware of the existing situation, which can be characterized in communication systems' terms as a 'one-way story-telling mode'.

Higher-levels of participation allow users to *interact* with the information, as well as retrieve and share knowledge. (Bugs et al., 2010) identified that a key aspect in collaborative geographic applications is the interoperability between geospatial data and tools available to Internet users wanting to build up their content. Depending on the level of participation, applications can vary from simple geospatial data visualisation portals to more interactive systems. Most of the applications can be found in the field of urban planning, e.g. London Profiler (2009), Argumentation Map (Rinner, Keßler et al. 2008), OpenStreet Map (2004). Those applications are straightforward for drawing something on a map, post descriptions and comments on specific buildings or areas, and the like. However, they cannot answer or analyse questions like: 'what is the influence of this new infrastructure on the surrounding environment?' Whether or not a system can answer these kinds of questions is the indicator of 'collaboration' or high-level participation. Because one needs to understand how decisions influence other participants, integrated modelling is one of the aspects in realizing collaboration. Accessing each of these models is another aspect of building collaborative applications. A high-level collaborative geographic application should include geospatial information representation, data management, scientific modelling, and smooth interaction between users and applications.

6.3 Case Study: Coast management in Holland Coast

The Dutch coastline is about 350 km long along the south-east part of the North Sea. It is divided into three regions: the Delta coast in the south, the Holland coast in the center and the Wadden coast in the north. The Holland coast between Hoek van Holland and Den Helder is typically a wave-dominated straight sandy coast. The sandy coast has the primary function of protecting the low-lying hinterland from flooding. Meanwhile, it is also important to other functions: e.g. ecological values, drinking water supply, recreation, residential and industrial functions. There are two main pressures along with the coast: (i) coastal erosion and (ii) socio-economic development. (15 year management dutch coastal) (Fig 6.1) Provincial committees are involved in stakeholder participation on coastal zone management and decide on the yearly national nourishment schemes. The committees are made up of representatives from relevant provincial authorities, water boards, municipalities and stakeholder organizations. Coastline protection is a crucial task in coastal zone management. The most important concern is the design and execution of a yearly nourishment program to distribute 12 million m^3 of sand along the Dutch coast. During meetings stakeholders can make remarks on the yearly nourishment program. This process requires a collaborative platform, which can support specialists to express their scientific considerations and assist the stakeholders (including general public) to illustrate their ideas.

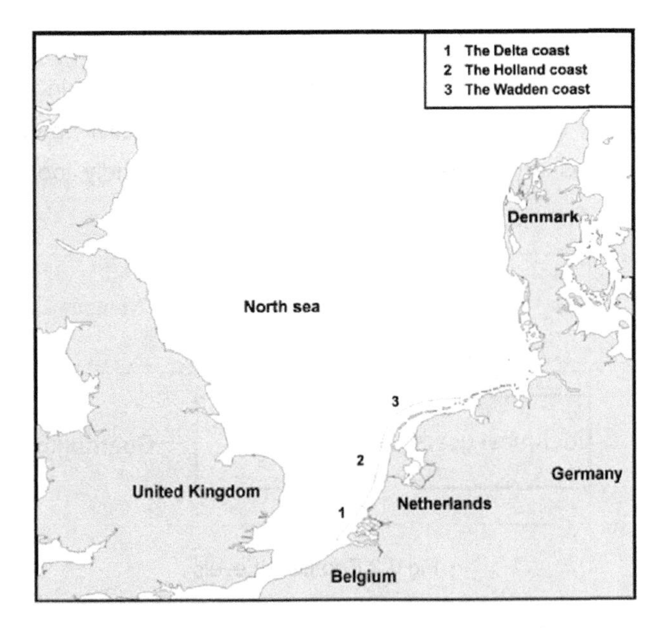

Fig 6.1 15 year management dutch coastal

Fig 6.2 3D view of Sand Engine Experiment, Kijkduin, South Holland

6.4 Prototype

6.4.1 Design considerations

In this case study, end-users were divided at two levels: (i) the contributor level, and (ii) the managerial level. (Fig 6.3) The contributor level explores alternatives by analyses. The managerial level refers to decision-making based on those alternatives. There are two groups of users at the contributor level: technical users (specialists) and non-technical users (stakeholders, citizens etc. who do not have specific technical knowledge). The design principle was to build a flexible platform to support all users in the discussion phase. The key was to find the balance between interactivity and visualisation capacities necessary to create an innovative user-friendly tool (Bugs et al., 2010).

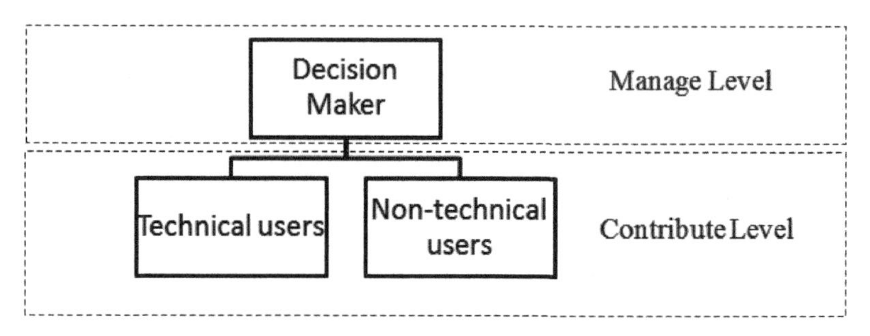

Fig 6.3 End User Levels

Spatial information is the most important in coastal management systems for both users. Maps should play a role as interface for both spatial referenced data and as platform for user to discuss of alternatives. Visualisation for both types of users refers to mechanisms of representing spatial data so they are unambiguously understood (Jo, Jason et al. 2007). Interactivity has different meaning to different user groups. Specialists intend to have high communication levels with the models running in the background and can easily adjust input parameters, execute models and retrieve the results in order to answer questions about what will happen in case of proposed measures. Being able to access models and parameters via the front-end are their main considerations. Non-experts normally prefer GIS tools and maps to better understand spatial effects of proposed projects, evaluate alternatives, and create new solutions (Jankowski, 2009). Using interactive techniques to identify interesting locations and define what should be done there, are their basic requirements. Fig 6.4 and Table 6.1 point out different actors in this system and their main tasks.

In current practice, joint projects commonly spend a significant part of their budget to set up some basic infrastructure for data and knowledge management. Most of these efforts disappear again once the project is finished (Van Koningsveld, 2010). Different from an ad-hoc project approach is the more continuous approach to data and knowledge management as required at the 'managing level'. Access to data, models and tools are considered main components of a management process in almost all marine and costal engineering projects. However, they need a container for storage implemented in a logical way to allow coordination of functionalities, which can be achieved by developing a generic user interface for visualisation and interaction.

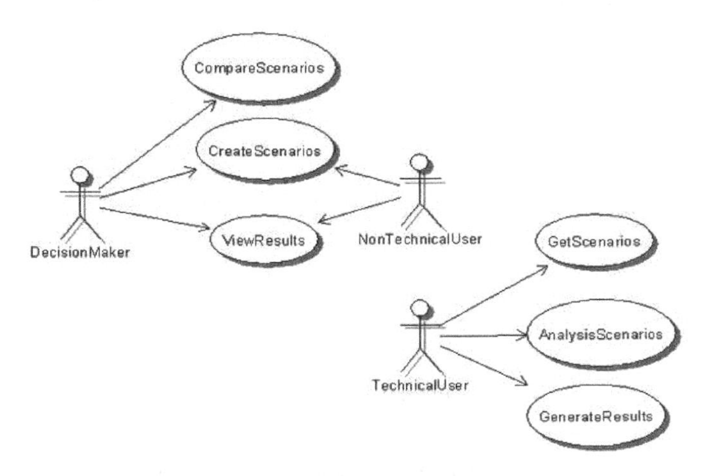

Fig 6.4 Use Case model for different users

Use Case	User Groups	Goal	Interactivities
CompareScenarios	DecisionMaker	Select the appropriate alternative	Viewing scenario layers and results layers simultaneously
CreateScenarios	DecisionMaker	Testing user preference scenarios	Add layers and information
	NonTechnicalUser		
ViewResults	DecisionMaker	Support decision making	Viewing results layers
	NonTechnicalUser	Understand the influences	
GetScenarios	TechnicalUser	Get input parameters from user interface	Information transfer though internet
AnalysisScenatios	TechnicalUser	Use the user created content for spatial analysis	Accessing model with user created content
GenerateResults	TechnicalUser	Visualize influence	Add layers with model results

Table 6.1 Analysis of Use Case model

Model-view-controller (MVC) is a computer software design pattern that separates the representation of information from the user's interaction with it (Trygve Reenskaug, 2009). A typical MVC structure is shown below. (Fig 6.5) The right part of the figure represents the knowledge extraction process via a generic user interface. The user interface should accomplish the components of 'view' and 'controller' as indicated in the left part. In this coast management case, 'view' has two layers: basic layer for file visualisation (2D and 3D), and tool layer for user-defined scenarios. The main responsibilities for 'Controller' are drawing functions and format transform function (from JSON to XML).

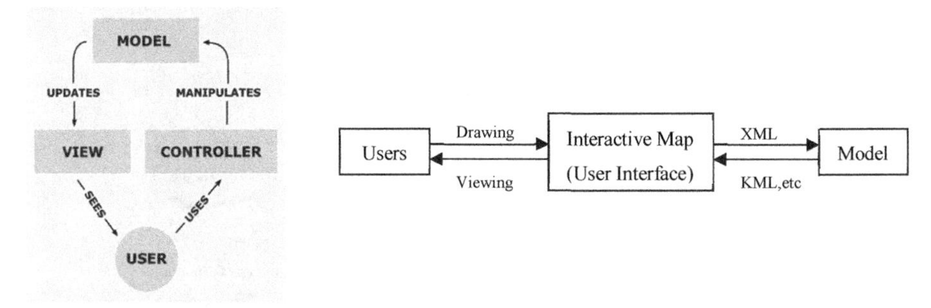

Fig 6.5 MVC pattern

The feature 'easy to extend' is particularly relevant at the managerial level. If the gap between a simulation model developer and a web interface developer can be bridged by a 'template' structure, this may facilitate the standardization process. Therefore, a highly abstract framework is required to realize such templates. Any 'template' can be considered a standardized style sheet that extends the interaction capabilities of a web page. Similar to the style sheet for graphical features in a webpage, a 'template' also standardizes the functions of an interface, in order to make it easier for non-web page developers to upload their own functionalities to the framework. The implementation of the 'template' is straightforward by changing the default function names in the scripts to the user defined function names.

6.4.2 Data source and viewer

Spatial data are basic to all kinds of users. The Netherlands probably has one of the best digital elevation models of the world: a 1m nation-wide grid coverage with the AHN DEM (Digital Elevation Map of the Netherlands) (Public Works and Water Management Geo-Information and ICT Department).

For the dynamic dune locations and foreshore beach dynamics, a time dependent DEM is available since 1996, presently at 5m-grid resolution. Moreover, all data is open to scientist for studying marine and coastal dynamics. There is a wealth of material to study the behavior of the Holland Coast and that can serve as an important resource for validating the predictive capabilities of simulation models. However, transferring these data into an appropriate format is a major effort for scientists who want to use them. There are numerous software packages to facilitate this process, but quite often these are hardly able to handle large amounts of data or the wide variety of available standards (Boer de, Baart et al. 2012). To policy maker and general public, this data are also most relevant. However, they often cannot make use of dedicated software packages used by marine and coastal scientists. More condensed visualisation tools are what they need.

Scientific models can be executed on the spot, therefore, the model results is one of the data source in this applications. Since the model results have been standardized into KML and XML format, the actual model layer is hidden to the end users. Meanwhile, in this architecture, various models can be extended as modules. For the prototype, the core of the model consists of the UNIBEST module that simulates coastline changes. The impacts on ecological indicators are calculated separately in post-processing modules, based on the outcomes of the UNIBEST model. These post-processing modules consider benthos, juvenile fish, and dunes (De Groot et al., 2012).

Web mapping sites are used by people that do not have specific GIS knowledge, but are willing to achieve spatial information. There are several standards for spatial data dissemination over the Internet, such as Web Mapping Service (WMS), Web Feature Service (WFS) and Web Coverage Service (WCS), which are issued by OGC consortium. However, the actual implementation of these protocols is lagging behind. Especially for time-dependent, 3D views of computational results on non-orthogonal grids, no standard implementations are available yet (Boer de, Baart et al. 2012).

KML is an open standard based on XML files certificated by OGC consortium. An advantage of KML is that it is supported by Google Earth (GE) which is familiar to different end-user groups. OpenEarth is a free and open source initiative to deal with Data, Models and Tools in marine and coastal engineering projects. With the OpenEarth community (Van Koningsveld 2010), a generic and coherent toolbox has been created for many atomic plot types using KML. Within this toolbox, , specialists have solved issues on coordinate system definitions. KML is also used in the GE environment to support full-featured visualisation.

Viewers need to assist end-users in creating scenarios. Unfortunately, the drawing functions are not so developed in GE for end-users to be able to input parameters and modify geometrical attributes. Quite often the user input scenarios are linked to simulation models which means data transfer back to the model does not need to use KML since visualisation is not taken into consideration in this direction of data transfer. For the Holland Coast case, the most important information end-users need to identify are the location and volumes of beach nourishment. These can be considered spatial information with particular attributes. Therefore, user input becomes a string of coordinates and numbers. JSON (JavaScript Object Notation) is a lightweight data-interchange format that is often used for serializing and transmitting structured data over the Internet as an alternative to XML. It can include spatial information into a geometry data type and attribute information into properties data type. JSON files (string data) are the main data structure for transferring user input to backend models.

6.5 Architecture and components

6.5.1 System description

'OpenViewer' is the name of the prototype developed to implement a generic framework for managing data, models and tools at a high-level. Fig 6.6 shows out the overall architecture of the prototype. The client side plays the role of controller and viewer. The Google Earth API is used to display the KML file in the database, whereas the OpenLayers APIs (http://openlayers.org/) in charge of the drawing functions for different tools. Extjs (http://www.sencha.com/products/extjs/) has been chosen as the layout framework for constructing user interface component. The JSON format is used as the standard format for representing spatial objects with non-spatial attributes. This fits nicely with client-side components implemented in most

JavaScript languages (Bugs et al., 2010). Moreover, it is easy to convert from vector layers to JSON files in OpenLayers. Moreover, the JSON format structure can be easily transformed to XML format, used for the work flow defined before.

At the server side, data transformation services can be considered as the 'wrapper' of scientific models. It receives the XML file from client side and, depending on the model requirements, abstracts the information and transfers to the model. After the model runs, it returns the results via KML files from the server to the web browser. One important detail is that the model should present its status to the client to describe each step, also by XML file. Then, the web page can pick up the actual status in order to tell users what is happening in the model and show a percentage mark of how much work has already been done. In this way, users can know the task is executing rather than receiving no information about what is going on.

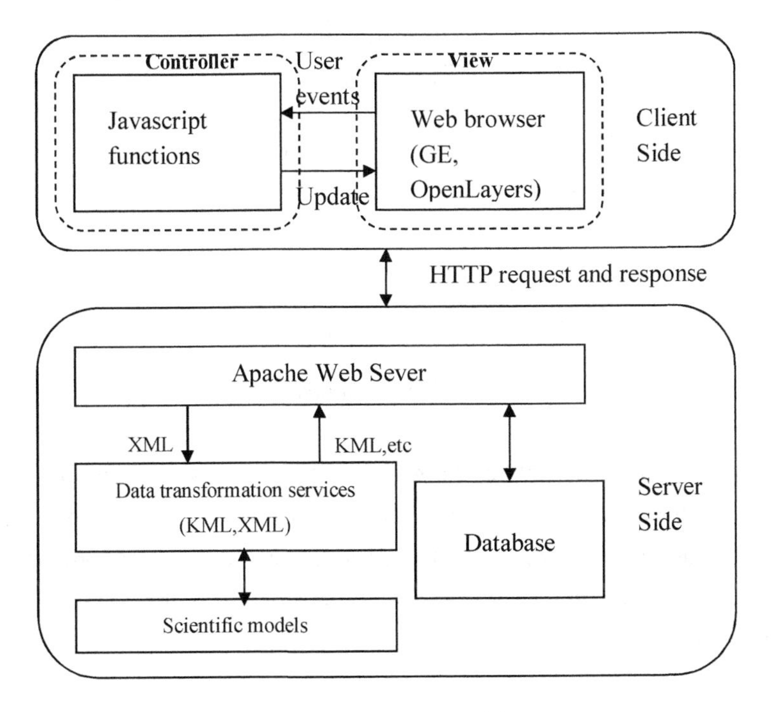

Fig 6.6 Overall architecture of the prototype

The GUI part of viewer has been composed by two parts: control panel (left) and map window (right). (Fig 6.7) The tree structure of the control panel is built by xml files. It is easy to modify by developers for adding new projects. The structure's hierarchy is shown in Fig 6.8.

Fig 6.7 Overview of Viewer user interface

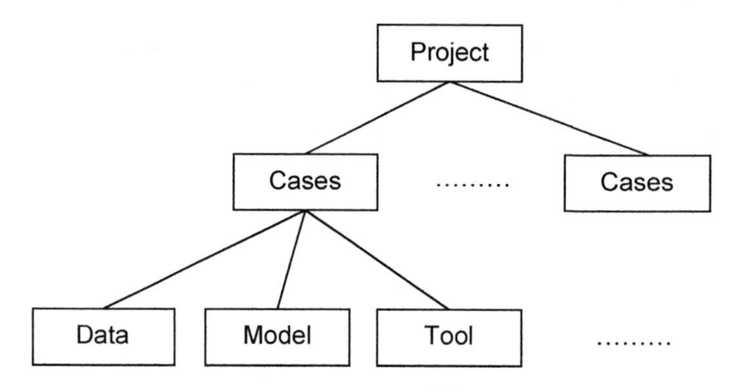

Fig 6.8 Project Structure

When opening a project folder, there are different cases each in turn having three tree structures: data, models and tools. The data tree contains all information about the study area. By selecting particular data, an image layer will be pasted onto the earth surface for representing the location of the spatial information. (Fig 6.9)

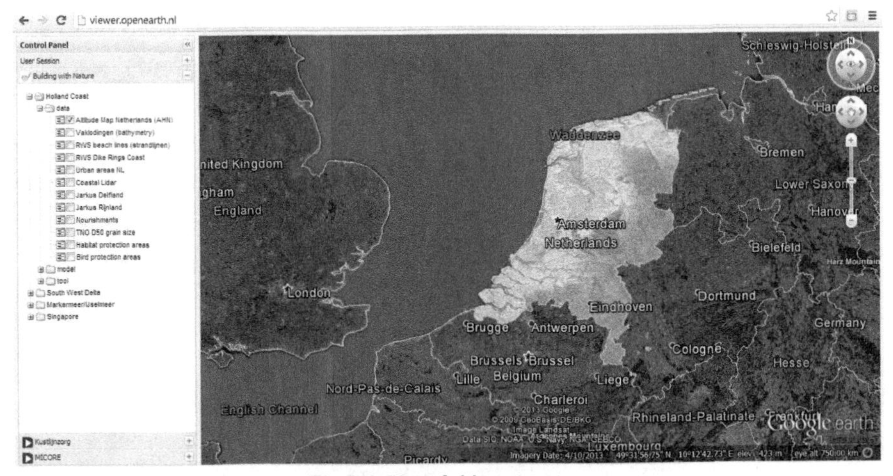

Fig 6.9 Data folder structure

In the model tree, users can select viewing the results of different models. For example, they can view the coastline position changes from 1965 to 2007 as indicated in Fig 6.10

Fig 6.10 coastline position changes from 1965 to 2007

Also, 2D and 3D model results can be visualized in the virtual earth window with more detail information. (Fig 6.11) (Fig 6.12)

Fig 6.11 2D model about Sand Engine

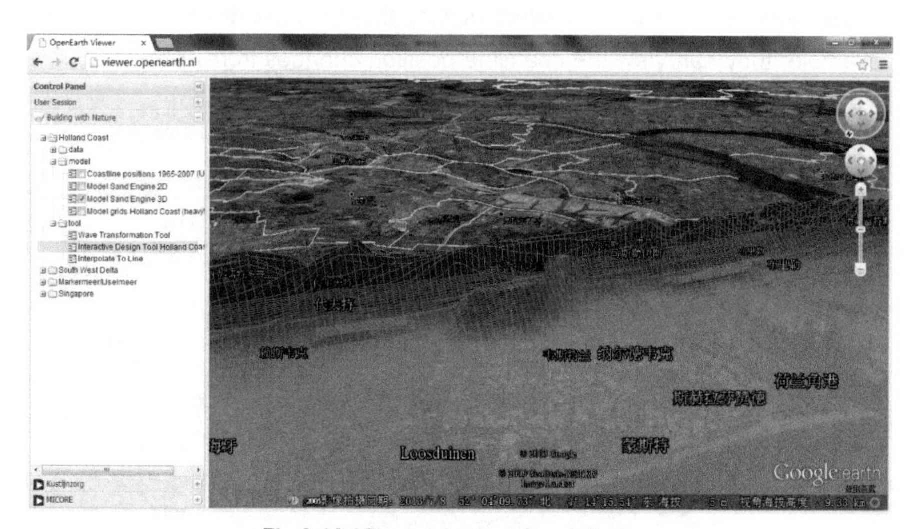

Fig 6.12 3D model about Sand Engine

The Tool tree includes different tools that can be used in interactive maps to execute online simulations. For instance, tools are available to derive the particular wave conditions at user defined locations. (Fig 6.13)

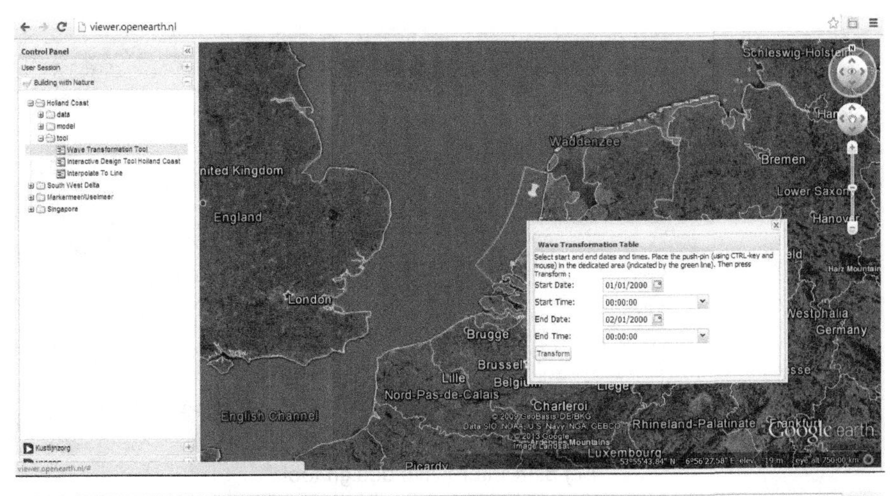

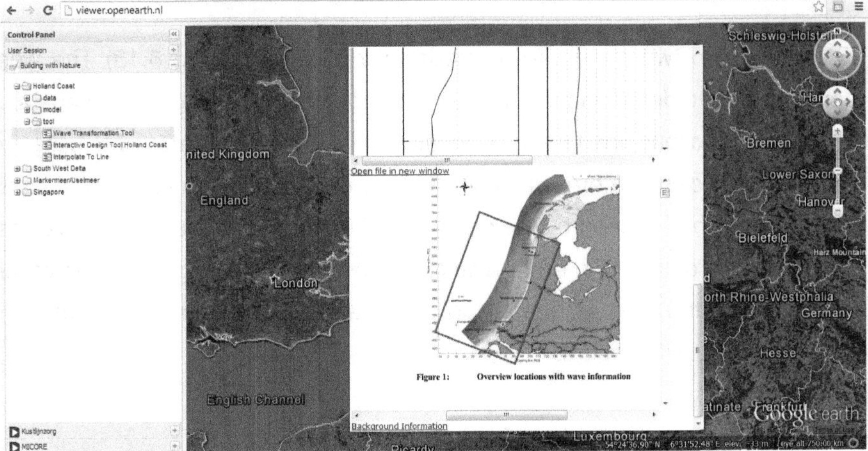

Fig 6.13 Wave Transformation Tool: User defined location(up); generate reports(down)

The interactive design tool can allow user to draw their own scenarios and send to the model in the server for online simulation. In this way, the what-if scenarios can be represented. (Fig 6.14)

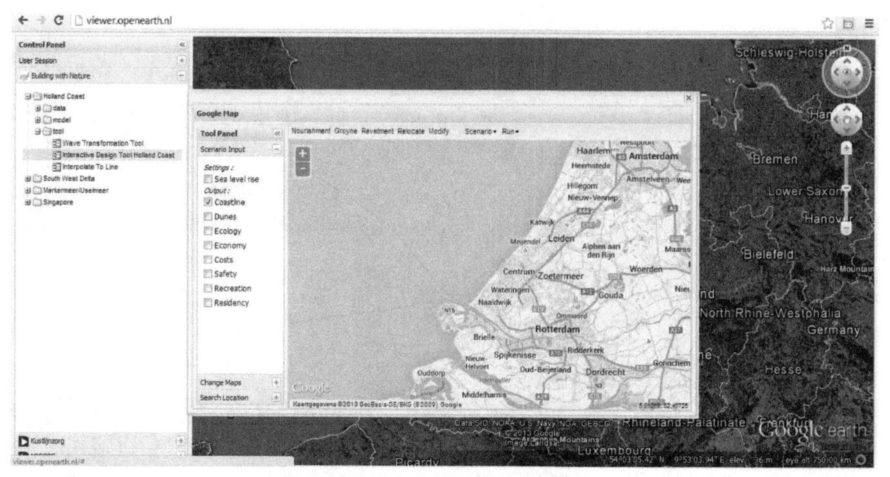

Fig 6.14 Interactive design tool

There are also two panels for composing tool windows.(Fig 6.15) The west panel is used to define the input settings, output layers, background maps and searching functions for finding a location that makes it easier for users to develop their own preferred measures. The east panel represents the study area which is pre-defined by the project xml file. The main features inside east panel are interactive map and menu for actions. By selecting the name of the measure that the user wants to add to the area, the mouse can be used to add the particular measure.

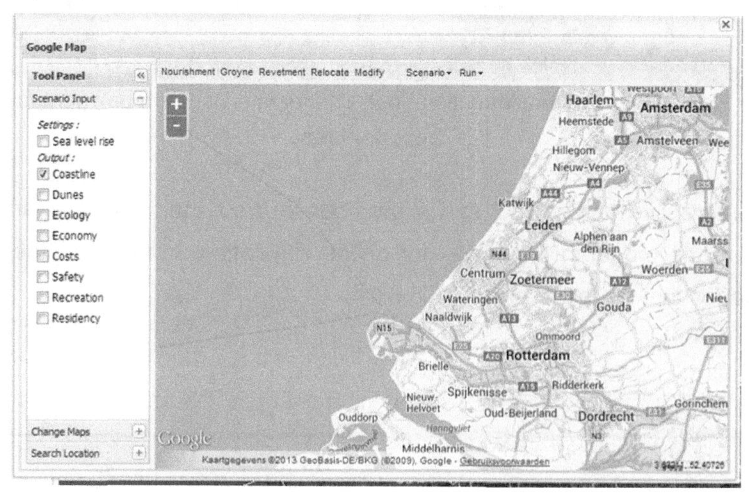

Fig 6.15 Interactive design tool window

A pop-up menu will appear to ask for more input on the measure that was just built by the user. (Fig 6.16) In this way, the user clearly understands that those attributes belong to the particular measure including the spatial location. In other words: Configurational knowledge is provided to the end-user.

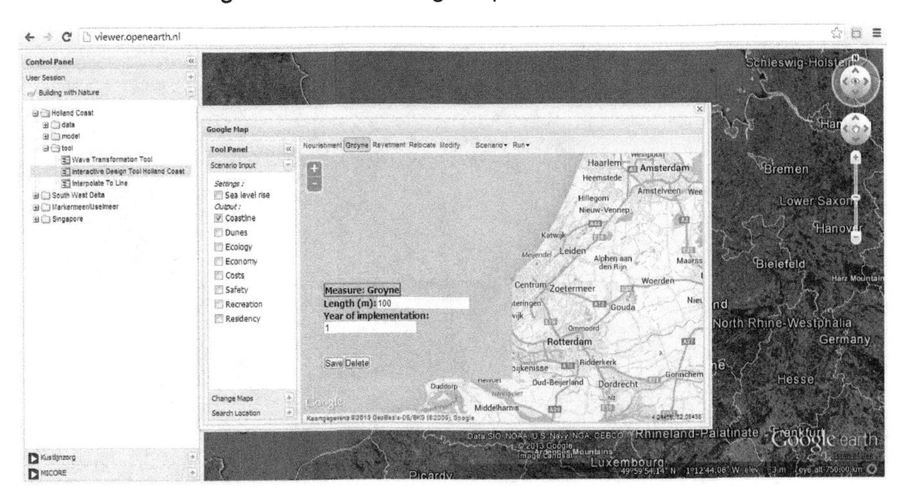

Fig 6.16 Pop-up for requiring input parameters

After creating user-defined scenarios, the input file is sent to the model for execution. Information messages are pop-up to inform users that model are running. From these message users can know the progress of their model runs, which will keep their attention to the webpage. (Fig 6.17)

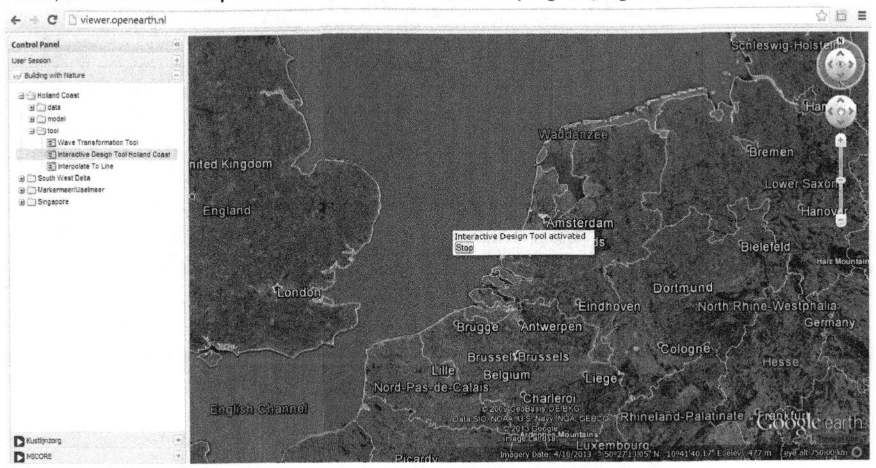

Fig 6.17 Back-end model has been executed

After the model runs are finished, the results are parsed into the KML files that can be projected onto the virtual earth window directly. At the same time, one extra panel will be shown for result management. Users can choose different information layers from the model results. Designed icons are representing different meanings of the model results.(Fig 6.18)

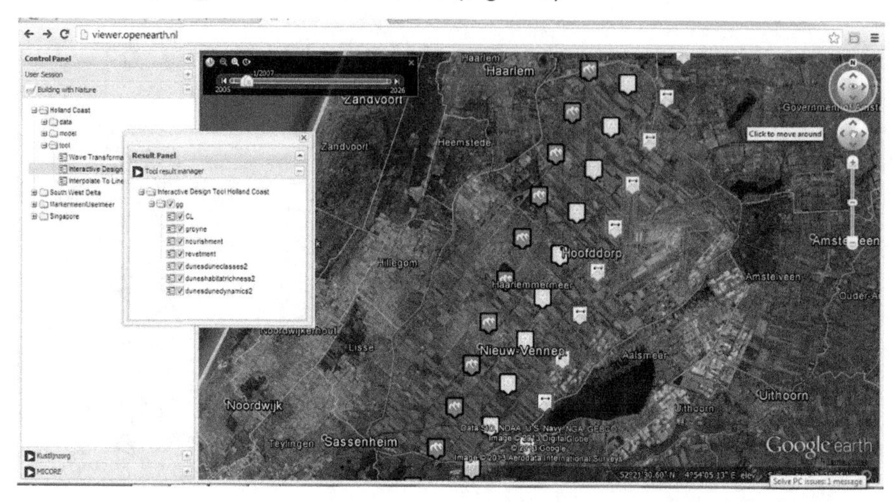

Fig 6.18 Model results has been sent back and parsed to KML layers

The interpolate tools can be used to view the altitude profile by user defined interpolate lines. (Fig 6.19)

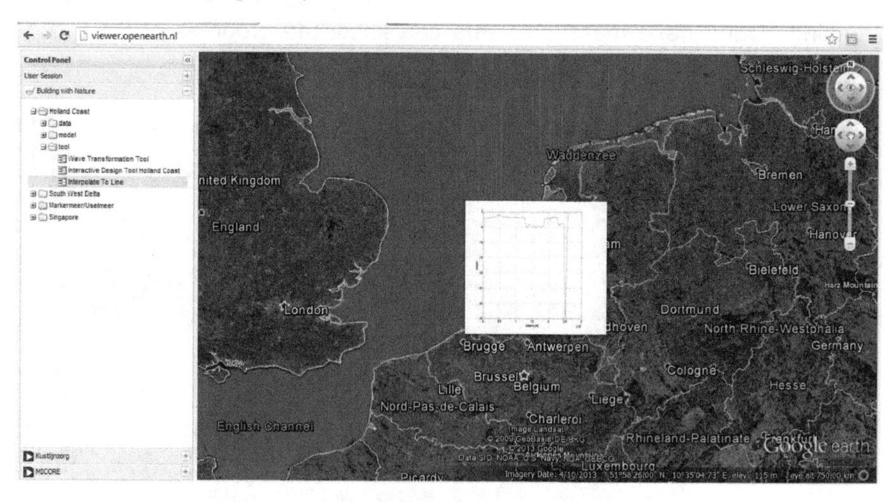

Fig 6.19 Interpolate Tool

Different study areas can be shift smoothly by clicking on the tree node which determines the areas. For example, Fig 6.20 shows that the viewport has been shifted to Singapore case folder.

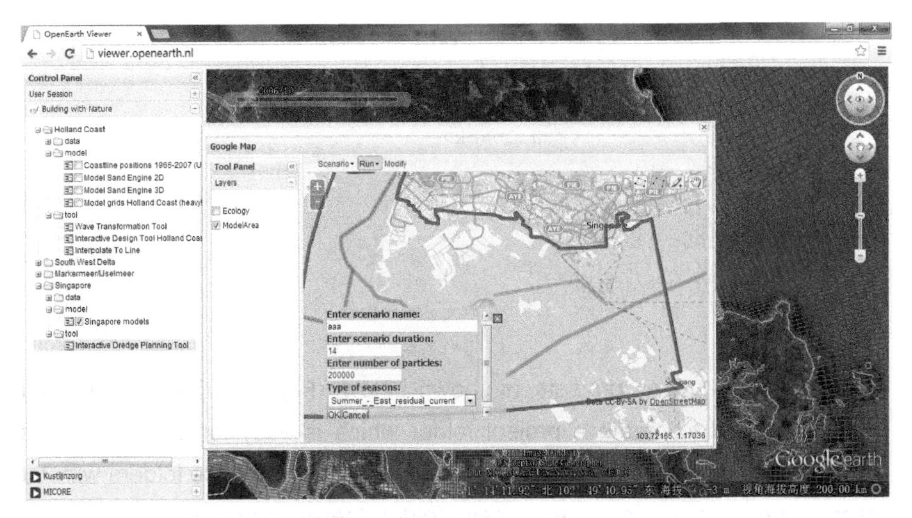

Fig 6.20 Singapore case folder

Regarding the 'Template' structure, the generic source code folder hierarchy is as indicated in Fig 6.21.

Fig 6.21 Structure of source code folder

The important folders are 'data' and 'projects'. There are xml files inside the 'data' folder which can be used to construct the tree structure in the control panel. When new tree node be added into the panel, developers only need to upload the corresponding xml files into this folder for generating the new node. The template for adding new tool into BwN project and HK case is to add the following code into the '.../data/bwn.hk.tools.xml'

```
<tool>
    <title>Template</title>                //name of your tool
    <admin>xuan.zhu@deltares.nl</admin>
    <callback>Template()</callback>     //Initialize function name
    <enable>0</enable>
    <lat>52.05249</lat>
    <lon>4.58105</lon>
    <height>750000</height>
  </tool>
```

The subfolders under 'projects' is shown below. (Fig 6.22) The name of the project is used to create a project folder which is easy for developers and managers to arrange. Inside the project folder, there are case folders with the names of each case. Inside each case folder, there are the tool folders, which represent the various tools for one case. Inside the tool folder there are the 'MVC' structure subfolders controls, gui and main. The 'controls' and 'gui' subfolders correspond to the 'controller' and 'viewer'. The 'main' folder is the initialization for each tool. The 'Model' folders are hidden on the server side. In the 'gui' folder, two scripts are in charge of constructing the window of this specific tool. In 'default' the map is provided as the background for each 'tool' window. The 'mapControl.js' is for controlling the activities on the maps. 'Popup' is an important feature for this application. After the users have added their own features to the map, popups are used to input the properties about these features. Developers only need to change the name of the template folder into their own tool name and replace all function names inside the different scripts which include the word 'template' into their own tool name. The basic functionalities of extended tools can then be visualized in the web-based viewer.

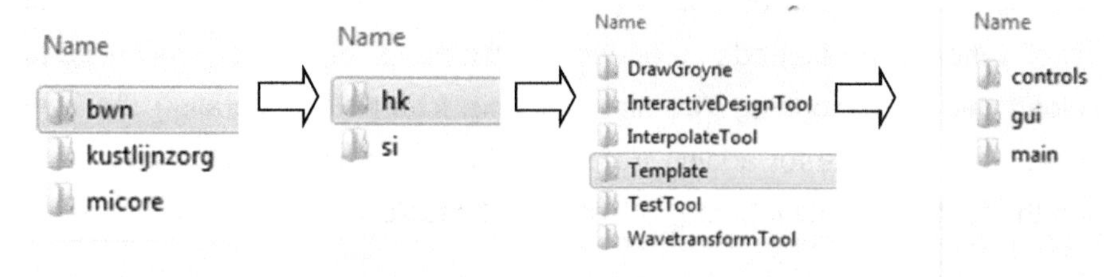

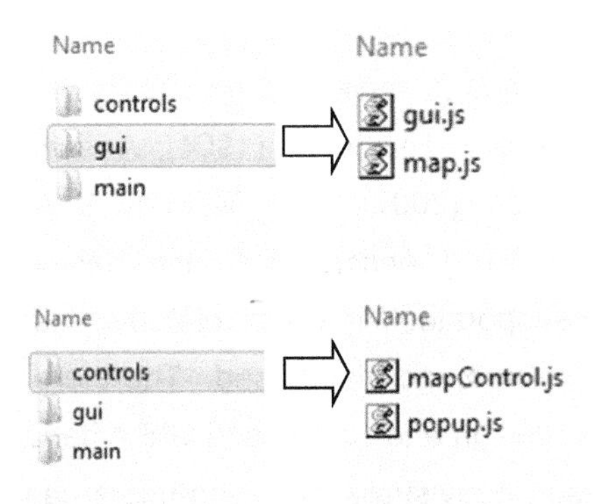

Fig 6.22 Structure of sub-folder system

6.5.2 Evaluation test

As mentioned already, user requirements play a dominant role in visualisation and interaction design. Hence, any evaluation should also be user-centered in order to test whether or not the design is in accordance with the requirements. Usability tests are appropriate to evaluate the quality of the framework design. Norman (1988) pointed out that no matter what you are designing, 'it should be brought back to everyday life' ('the design of everyday things'). Many disciplines (for instance, mathematics) do not always relate to everyday practice; in some fields of some computer science it may also be hard to immediately see 'the value of visualisation' (van wijk, 2005). Visualisation and interaction are tools that are used by people. Especially, in the information technology age of Internet, visualisation and interaction examples are easier to achieve in our daily life. The judgment on the quality of visualisation and interaction is in the end-users' hand.

Usability is defined in the ISO 9241 standard as "the effectiveness, efficiency, and satisfaction with which specified users achieve specified goals in particular environments". (Nivala et al. 2007) Effectiveness is defined as 'the accuracy and completeness with which users achieve certain goals'. Efficiency is 'the relation between the accuracy and completeness with which users achieve certain goals and the resources spent on achieving them'. This can be measured by *task completion time and learning time.* Satisfaction is the users' comfort and positive attitude towards the use of the system, measured by *attitude rating scales.*

Much work has been done in seeking guidelines to implement a usability test in the Human Computer Interaction discipline. (Nielsen, J. (1994); Artemis Skarlatidou, M. H. (2006); C.L.Sidlar, C.Rinner. (2007); Zafiri, M. M. H. a. A. (2008); Oztekin, A., A. Nikov, et al. (2009);) According to the design framework used in this thesis for decision support tools in water related areas, the majority of the information is spatially-related and map-based. Therefore, it is appropriate to apply the user interface design and web-based GIS criteria for evaluating their usability. For web-based geovisualisation, end-users can be divided into two groups: experts who regularly use geovisualisation to explore spatial patterns in meta geo-data sets – and non-experts, who don't. The methodology of the tests developed in this thesis is based on an understanding of several knowledge discovery activities, visualisation operations, and a number of steps in computational analysis used to visualize patterns in data (Koua et al., 2006). Since the knowledge level of non-experts can be diverse, simple to use and easy to understand tests are required. For the public at large, visualisation is used to present information and here the Public Participant GIS (PPGIS) is the result of emerging GIS appreciation by the public (Nyerges et al., 1997).

6.5.3 Evaluation criteria

Nielsen (1994) gave the heuristic evaluation to the user interface design. Sidlar and Rinner (2007) employed a quasi-naturalistic case study on the usability test for Argumentation Map. Zhao and Coleman (2007) performed a guideline that includes cost entry as one of evaluation criterias. Bugs (2010) employed an assessment for PPGIS in urban planning practice in Canela by using questionnaire-based interview to achieve users satisfaction. Within our framework, the purpose is to assist experts explain consequences of alternatives and help non-experts to represent their preferences. Therefore, the criteria should take both types of requirements into account.

1. Cost of entry: the price of the tools and components needed to run or access the prototype.
2. Ease of use: simple to understand and use the functionalities of the platform. For experts user/tool developer, the framework needs to be as simple as possible for them to link their model to the front end

component. For non-expert users, fast access to the model, consistent navigation through the database and 'what-if' scenario representations are key factors influencing the sense of 'easy to use'.

3. Effectiveness: the proficiency of achieving the tasks by using the tool. It is important to make sure that the information is transferred correctly by the visualisation and interaction techniques.
4. Efficiency: the time spent on completing each task. It measures the performance of the system and strongly related to the 'easy to use' criteria.
5. Satisfaction: users' altitude during the whole testing phase. Users can finish all the tasks and give all the correct answers within a reasonable time (especially expert users), but may still have reasons that retain them to use a tool. It is important for web-based applications to support content-based thinking, viz. thinking in terms of the actual model outcomes, not about the procedure of how to obtain them.

In order to acquire end-users' feedback about the web application, two surveys were held, one during the prototyping phase and one right after user tasks test. The first questionnaire was used to collect the end-users' expectations, the second to measure their feedback after using the application. The outcomes are presented below.

6.5.4 Practical strategy

To evaluate the OpenViewer prototype, three activities were organized. First an introduction session was held during the coastal management workshop (Atelier Kustwaliteit) on June 1st 2012 in Driebergen, the Netherlands. This was the first time the prototype design was shown to a wider public of potential end-users. It has been developed for use during stakeholder sessions, for rapidly giving a first impression of the consequences of envisaged management scenarios. During this workshop, a presentation about the OpenViewer concept was given and there was a session where questionnaires were distributed to participants for collecting their ideas about the presented framework at the beginning of the design phase.

The second evaluation was held on October 19th 2012 at UNESCO-IHE after the prototype had been developed, where end-users were asked their opinion about working with the tool. During this evaluation, users were asked to evaluate three components of the system to test their understanding about using the OpenViewer. The first aspect was related to the 'data' part of the platform where users were to choose data from the data folder and then asked how they perceived the dataset. The second and third aspect were on how to use the tools made available in the 'tool' folder, and questions were asked to test whether or not end-users can understand the model results by themselves. Before the test, no explanation was given about how to use the tool, in order to check the users' first impression about the application.

The third evaluation was among the developers in the team itself, testing the ease-of-extension functions in the 'Template' concept. Together with three other developers, the author worked for two months on ways how to extend the tools from the main web page.

6.5.5 Results

- First evaluation

Fourteen feedbacks were received from participants of the first session. The average age was about 47 years old and participants were coming from a wide range of organizations with different background. From the results of the questionnaires it was observed that all of the participants thought that Google Earth/Maps would be useful as a visualisation platform. Some 80% of the participants were willing to accept several minutes of waiting for the modelling results. All participants agreed that the web-based interactive applications presented in the workshop were useful; almost half even said it was crucial. In order to judge coastal line change, most of the participants wanted to view the cost as the reference in the application, almost half wanted to view the dune development, and many were interested to see the ecological effect.

- Second evaluation

After the actual application had been developed, a small workshop was held at UNESCO-IHE attended by 8 participants and focusing on end-user task evaluation. The average age of the participants was about 28 years. There was no training or explanation session at the beginning of the test. Most of the participants found the application easy to use and adequate to evaluate alternative scenarios for coastal interventions. Almost all agreed that the platform was easy to teach other people the basic principles on the effects of coastal interventions. More than half of the participants liked the interaction method of the platform and would like to use it more often. Almost all thought it could provide enough information for decision making, and could be used either independently or during expert explanation.

- Third evaluation

The third evaluation was to check whether or not this platform is easy to extend for other tools and easy to maintain. The three 'new' developers involved in this evaluation all had their background in modelling applications rather than in software development. During the two months of working together, the new developers quickly understood the structure of the platform. It was easy for them to create new map-based interactive tools. They could easily focus on defining new user actions with the help of the tools and could by themselves design the interfaces to the corresponding models

6.5.6 Discussions

From the usability tests it was found that the majority of the participants considered the platform to be useful for project management, modelling, education, decision-making and discussion. The background knowledge for getting started using the Google Earth/Maps features are widely favored by the majority end-users. During joint meetings with experts, public participants found it easy to create their own future scenarios and ask experts to explain the model results. The tool is most suited in the beginning of the design process, when all possible scenarios are being explored. The model should not be used without moderation by an expert, who can clarify the results if necessary and explain model limitations. In this way, domain knowledge can be incorporated earlier in the decision making process, which was the objective of the research presented here.

6.6 Summary

Making advanced numerical modelling systems publically available online, using easy to interpret GIS systems or appealing graphics visualisation techniques, is not without danger of mis-use or mis-interpretation. Most often experts are needed to explain the outcomes of different scenarios, as in the case of coastal zone management discussed in this chapter. However, visualisation and interaction can greatly contribute to better understanding the consequences of proposed measures, not only to experts but also to all stakeholders involved, and even to the public at large.

The experts' task is to make sure the simulation models are working as correctly as possible, whereas achieving an appropriate decision is the decision makers or end-users task. Mutual trust is a necessary condition. From this condition, experts can create a virtual reality and decision makers can create their preferred solution. Web-based techniques enable easy access to data, models and tools. Creating a virtual environment can assist end-users in defining their preference. Using figures, diagrams and computer animations can help to convey the effects of proposed measures and share relevant knowledge. In the case of coastal zone management addressed in this chapter, virtual environments based on geo-graphical information systems and visualisation of numerical model simulations are seem to be essential for displaying the effects of proposed measures and for obtaining end-users' feedback in case of public participant decision making.

7 Conclusions, Recommendations and Outlook

7.1 Conclusions

The objective of this research was to demonstrate the potential of computer-based visualisation and interaction techniques in decision support systems for practice or modelling frameworks for research. The value of visualisation and interaction in decision support tools depends on the different user requirements that can be distinguished in the decision making process and on using corresponding visualisation techniques to achieve the appropriate user interface. The need for implementing visualisation and interaction techniques also comes from increasing the cognitive level in the decision-making process, especially in the alternative selection phase. In this research, a disaster management system was used to analyse the user requirements and recommend corresponding visualisation techniques. Many decision support systems are sharing the same user interface as the underlying simulation model software. This leads to the problem of complex options, difficult interaction, different interpretation and time-consuming efforts for stakeholders user groups in the decision support phase. A distinction should be made between end-users and model-developers. An Object Oriented (O-O) approach can be used to classify the various groups of end-users.

Using a relatively simple to interpret 2D web-based map interface has the advantage of allowing a range of different stakeholders from different background areas to participate in the decision-making process. Such approach can be widely used during the risk management phase that requires expert knowledge (on modelling) to assist decision makers. The link between the underlying scientific models and the end-user interface is important for satisfying both easy use and quick access. Surrogate models proved very valuable for solving time consuming simulation runs in case of process-based modelling. Clickable Items and Layered Control are found to be two of the most important factors in UI design. By clicking on the spatial features on a map, different measures can be implemented onto specific features. In case of flood modelling, upstream and downstream relationships are represented explicitly to help the user understand more about the impact of proposed measures on the river basin as a whole. techniques are important features in Web 2.0 that can help generate a smooth user experience for easy-to-use web based applications. Evaluations show the compatibility of those techniques and lightweight front-ends for on-line applications. The framework of the VisREACHER tool in Chapter 3 is such example that can be useful for decision makers in water resources management and also for potential users who require an easy-to-use tool for understanding the impact of different measures.

Creating a Virtual Environment is another important spatial information representation method. With 3D graphics display, users can obtain a direct feeling about the surrounding situation. This is useful in information dissemination and communication, especially for public participation sessions. Two main components can be distinguished in virtual environments: (i) basic objects and (ii) event objects. Digital Terrain data is one of the most important basic objects. DEM data are the main input data for constructing 3D terrain surfaces by generating a large number of terrain meshes. Methods for fast rendering those meshes can give a smooth vie and feeling of interaction. The Level of detail (LoD) technique is a method to limit the number of terrain meshes to be rendered for each view. The Node Evaluation System used in this thesis proved key to judge when to reduce the mesh. Combining distance between viewpoint and terrain node with the roughness properties of the terrain node itself showed to be an efficient way to constitute an evaluation system and appeared easy to implement in algorithmic form. Event objects can be considered as the surface information that is merged with terrain. For flood disaster management systems, the water surface plays a crucial role for constructing event objects. Simulation models present water levels as time series, which can be used as the input data source for making a flat water surface which is convenient for simulating a flowing river surface (the slope of the free surface may only be 10^{-4}).

In real world applications the question how to include 3D information into existing disaster management systems is strongly related to how to describe the process of disaster management. Flood disaster management systems were considered here as an example to explain how to determine the various phases and actors and relationships between them. It may be worthwhile to point out that application of 3D data seems preferable for sending flood-warning messages and for emergency response. Semantic information about 3D objects also needs to be revealed by combining 3D and attribution data types.

Bridging the knowledge gap between different stakeholders is a separate area of expertise. How to bring expert knowledge earlier into the decision making process is important to assist decision makers and public participants in understanding the consequences of their decisions to other stakeholders. In this thesis, web-based techniques were combined with advanced user interface design to help define the appropriate way for end-users to interact with different data sets and scenarios. Parameter settings should be filtered to make sure that end-users understand the meaning of different settings. Complex model parameters should be hiden in the public user interface. Using platforms that end users are familiar with can make DSS features and functionalities easier to understand. Displaying the system status information in the user interface is important in particular during runtime of the underlying simulation models, which could take quite some simulation time. If no information is made available on what is happening, users may terminate the process by mistakenly thinking that there is a system error. Using a general structure of component programming which works like 'template' can fill gaps between GUI-developers and model-developers. In this way, component-model developers can focus on content and specific process formulations, rather than on user interface design or user logic.

7.2 Recommendations

Decision support systems usually target end-users that are not model-developers and therefore require a user-interface different from expert users, in order not to become confused by the numerous options in various modell settings. Before developping any decision support tool, understanding the user groups' needs is necessary for identifying user interaction. It is recommended to use Object-Oriented methods in the phase of defining classes of users and their affiliated user-requirements.

Both 2D and 3D visualisation techniques are useful for representing spatial information. A 2D map is good at describing Declarative knowledge and is suitable for representing scenarios for general situations. However, when Procedural and Configurational knowledge is required e.g. in an emergency response phase, it is recommended to use a 3D virtual environment since this is capable of better / faster representing the on-site situation with more detail than a conventional 2D map.

Web-based decision support tools can disseminate information more widely and support more people taking part in the discussion process. Web 2.0 can provide a flexible and easy to use browsing environment, which can collect information and interact with models. Easy-to-use menu's, layered control structures, legend bars and status bars are basic components for user interface design and readily available in Web 2.0. However, click and select features on a map-based interface need to be designed carefully in order to obtain the required information in a limited number of clicks.

Pop-ups are very useful as a tool for obtaining input parameters. It also saves space on the webpage and can enlarge the viewport for maps. If time-consuming models run on a server, it is recommended to deploy a real-time status bar on the front-end to give the end-user some general information about expected model execution time.

Using surrogate models is an option to reduce execution time of complex modelling systems (e.g. the Bayesian Belief Network introduced in this thesis as a proxy for a more complex spatially distributed simulation model). It is recommended to implement such models in a web-based modelling environment, provided the accuracy needs are taken into account.

It is recommended to include end-user testing results as one of the evaluation criteria for web-based decision support systems: effectiveness, efficiency *and user-satisfaction* are three main factors in evaluating modern-day software systems.

7.3 Outlook

Advanced computer graphic methods and visualisation techniques bring us a new view angle to obverse our world. Computer-Aided Virtual Environment (CAVE) techniques may have long been used in e.g. pilot training, but they are not commonplace as a workbench for disaster management or for decision making in water resources management. But also in these areas, advanced visualisation methods and interaction techniques can find powerful applications. Flood Early Warning Systems (FEWS) and disaster management are just two examples. In the past, advanced immersed visualisation required costly hardware devices and software packages that needed to be maintained by highly skilled professionals. But more recently, affordable web-based visualisation technologies are becoming available that can be used in research as well as in practice, like decision support systems for water related problems.

Obviously, a detailed 3D representation of e.g. a flood study area can display specific features like flood propagation speed and water levels. This is very useful in flood control rooms for decision makers to assess dangerous situations and make appropriate decisions well ahead of time. However, if stakeholders or decision makers are not all in the same location, a flexible method is needed that can share images and communicate the effects of potential measures. For example, the requirement from on-site actors in emergency response may just be to have a simple flood map prediction. The *value* of visualisation – in itself merely a technology, although sometimes a very impressive one – is to be measured primarily by its usefulness.

This is not always recognized in the visualisation research community, where individual researchers are often driven by increasing their number of publications on which they are judged as academics, and not on implementing their methods in valuable practical systems ready for end-users (Wijk, 2001). It would be wise to also consider practical usability and to identify end-user requirements for guiding research on appropriate methods for visualisation and interaction. The field of 3D-GIS is such discipline dealing with storing and retrieving spatial information for dissemination into an easily comprehensible format, even for non-visualisation experts.

Spatial database management and 3D modelling play an important role in several areas of geography like urban planning or flood control. Advanced 3D city models help research the best ways to achieve sustainable urban development. Also, disaster management systems often rely heavily on having detailed 3D spatial information quickly at hand. Two phases are commonly distinguished here: *risk management* and *emergency response*. The main actors are different in each of those two phases, and so are the user requirements. Time is more critical in emergency response where simple but easy to understand visualisation methods can suffice for developing on-site scenarios. In risk management time is often less of a constraint, so more advanced but time-consuming display methods can be used in this case.

Web-based technologies are widely used for transmitting data and disseminating information. Online data format standards like e.g. CityGML have been created especially for arranging complex structures in 3D cities in order to facilitate carrying out spatial analyses and performing data mining and thematic inquiries. Abstraction is an important concept that translates processes into a computational language that then can be digitalized for further management purpose. The Unified Modelling Language (UML) is a standardized general-purpose modelling language in the field of object-oriented software engineering. By using this tool, static and dynamic processes can be transformed into various diagrams that can easier display the logical structure for software design and implementation.

Let us take the development of the decision support system for the Netherland Institute for Safety (NIFV) as an example, where the original goal was to gather all experts working in the same place to work with one of the Command and Control (C&C) systems currently available in the Netherlands (Zlatanova, 2008). Then, web-technologies for risk management and emergency response were introduced and, as an exercise, several teams were asked to resolve a hypothetical dyke break situation in Flevoland, a polder in the central Netherlands. Every team was composed of experts on traffic and transportation, on flood modelling and on demographic data evaluation. The existing C&C system had many GIS data organized in different layers. The flood modelling simulations provided information on future flood scenarios in time steps of four hours. Experts were asked to find out who were in the threatened area and how to evacuate them. Every hour, they had to report to the mayor and the press (represented by the instructors). The software they used for analysing the spatial information and generating thematic maps was ESRI's ArcGIS.

The user interface was clear even to non-expert users, for navigating through the layered structures, but not easy to implement in the analysis toolbox for non-GIS specialists. Collaboration was key in this experiment, because emergency response is teamwork that involves different stakeholders. Web-based GIS techniques proved to have the greatest potential for facilitating collaboration. From the exploration of 3D-GIS data to the implementation into a disaster management system, online map-based visualisation was able to achieve a balance between detailed construction rendering and transiting useful information. Spatial databases and online map services play an important role in this balance. Several supercomputing centres are taking this into consideration for visualizing scientific data and for communication, e.g. SARA (www.sara.nl) in the Netherlands.

SARA is one of the world's largest scientific computational centres that support research in the Netherlands by developing and offering an advanced ICT infrastructure, services and expertise. A recent example is the Collaboratorium, a new facility that directly tailors high-resolution displays, remote data rendering, fast user-interaction, cooperation and user friendliness to the current and future needs of scientific and business communities. Application areas range from developing new medicine (Van der Spek, 2012), managing floods (Belleman, 2012) to investigating bird migration behaviour and performing virtual fieldwork in ecology (Bouten, 2012). The latter combined web-based techniques with GPS measurements and data visualisation techniques in a 3D-GIS virtual environment. The result could be viewed from any normal computer with a Google Earth plug-in. The display shows the fly-paths of the birds with the landscape map as the background. It is as if one 'flew with the birds' oneself. Such combination of dynamic and static spatial data enables bird-researchers to analyse and understand bird behaviour and migration processes. Since only low-level display devices are needed, this gives the flexibility of entering the data anytime from anywhere. Free and Open Source Software (FOSS) becomes a valuable tool in such custom tailored systems. Standardized data formats are important for developing web-based systems that can facilitate client-side and server-side information transformation.

References

Ahmad, K. a. P., R K (1996). "Safe Hydroinformatics NATO Advanced Study Insititute on Hydroinformatics Tools for Planning, Design, Operation and Rehabilitation of Sewer Systems", Harrachov, Czech Republic.

Aldescu, G. C. (2008). "The necessity of flood risk maps on Timis River". IOP Conference Series: Earth and Environmental Science, IOP Publishing.

Alter, S. (1980). "Decision support systems". Reading Addison-Wesley.

Arens, C., J. Stoter, et al. (2005). "Modelling 3D spatial objects in a geo-DBMS using a 3D primitive." Computers & Geosciences 31(2): 165-177.

Argent, R. M., A. Voinov, et al. (2006). "Comparing modelling frameworks â€“ A workshop approach." Environmental Modelling & Software 21(7): 895-910.

Argote, L. and P. Ingram (2000). "Knowledge transfer: A basis for competitive advantage in firms." Organizational behavior and human decision processes 82(1): 150-169.

Arta Dilo, S. Z. (2008). "Spatiotemporal Data Modeling for Disaster Management in the Netherlands". the Joint ISCRAM-CHINA and GI4DM Conference, Beijing.

Artemis Skarlatidou, M. H. (2006). "PUBLIC WEB MAPPING: PRELIMINARY USABILITY EVALUATION". Proceedings of GIS Research UK Conference Nottingham.

B. Eckman, P. C. W., C. Barford, G. Raber (2009). "Intuitive simulation, querying, and visualisation for river basin policy and management " IBM Journal of International Business Machine 53: 7-18.

Bacon, P. J., J. D. Cain, et al. (2002). "Belief network models of land manager decisions and land use change." Journal of Environmental Management 65(1): 1-23.

Bas Pedroli, M. M., and Michiel van Eupen (2007). "Yellow River Delta" Environment Flow Study.

Basic, F., Handmer,J. and Cartwright,W. (2005). "An Evaluation of the Flood Warning Information System". MODSIM 2005 International Congress on Modelling and Simulation: 2790-2795.

Berlekamp, J. r., S. Lautenbach, et al. (2007). "Integration of MONERIS and GREAT-ER in the decision support system for the German Elbe river basin." Environmental Modelling & Software 22(2): 239-247.

Bezuyen, M. J. "Flood management in the Netherlands."

Bhargava, H. K., D. J. Power, et al. (2007). "Progress in Web-based decision support technologies." Decision Support Systems 43(4): 1083-1095.

Bishop, I. D., R. Bruce Hull Iv, et al. (2005). "Supporting personal world-views in an envisioning system." Environmental Modelling & Software 20(12): 1459-1468.

Blake, I. (1980). "Review of "Decision Support Systems: Current Practice and Continuing Challenges, by Steven L. Alter", Addison-Wesley, 1980." SIGMIS Database 11(4): 23-24.

Blankenhorn, K. (2004). "A UML Profile for GUI Layout". Digital Media, University of Applied Sciences Furtwangen. Master.

Boer de, G. J., F. Baart, et al. (2012). "OpenEarth: using Google Earth as outreach for NCK's data".

Booth, N. L., E. J. Everman, et al. (2011). "A Web-Based Decision Support System for Assessing Regional Water-Quality Conditions and Management Actions." JAWRA Journal of the American Water Resources Association 47(5): 1136-1150.

Brown, D. G. (2006). "Agent-based models. Our Earth's Changing Land: An Encyclopedia of Land-Use and Land-Cover Change". H. Geist, Westport CT: Greenwood Publishing Group: 7-13.

Brown, I., S. Jude, et al. (2006). "Dynamic simulation and visualisation of coastal erosion." Computers, Environment and Urban Systems 30(6): 840-860.

Bruce, E. (1995). "Thinking in C++", Prentice-Hall, Inc.

Bugs, G., C. Granell, et al. (2010) "An assessment of Public Participation GIS and Web 2.0 technologies in urban planning practice in Canela, Brazil." Cities 27(3): 172-181.

Burkhard, R. A. (2004). "Learning from architects: the difference between knowledge visualisation and information visualisation". Information Visualisation, 2004. IV 2004. Proceedings. Eighth International Conference on, IEEE.

Ceri, S., F. Daniel, et al. (2006). "Modeling web applications reacting to user behaviors." Computer Networks 50(10): 1533-1546.

Charvat, K., P. Kubicek, et al. (2008). "Spatial data infrastructure and geovisualisation in emergency management". Resilience of Cities to Terrorist and other Threats, Springer: 443-473.

Chau, K.-W. (2007). "An ontology-based knowledge management system for flow and water quality modeling." Advances in Engineering Software 38(3): 172-181.

Chawla, S. S. a. S. (2003). "Spatial Databases: A Tour", Prentice Hall.

C.L.Sidlar, C.Rinner. (2007). "Analyzing the Usability of an Argumentation Map as a Participatory Spatial Decision Support Tool". URISA Journal 19(1): 47-55

Coad, P. and E. Yourdon (1990). "Object-oriented analysis", Yourdon Press.

Colin Starr, P. s. (2004). "An Introduction to Bayesian Belief Networks and their Applications to Land Operations", Land Operations Division System Sciences Laboratory
COST Training school report (2009). "3D Geo-information for Disaster Management".
Cumming, G. and C. Norwood "The Community Voice Method: Using participatory research and filmmaking to foster dialog about changing landscapes." Landscape and Urban Planning 105(4): 434-444.
D.P.Loucks (2005). Water Resource Planning and Management, UNESCO.
Daniel, T. C. and M. M. Meitner (2001). "Representational validity of landscape visualisations: the effects of graphical realism on perceived scenic beauty of forest vistas. " Journal of Environmental Psychology 21(1): 61-72.
Danny Poo, D. K., Swarnalatha Ashok (2007). "Object-Oriented Programming and Java", Springer London.
Dave, C., P. Eric, et al. (2005)." Ajax in Action", Manning Publications Co.
David, C. W., L. Heather Richter, et al. (2008). "Charting new ground: modeling user behavior in interactive geovisualisation". Proceedings of the 16th ACM SIGSPATIAL international conference on Advances in geographic information systems. Irvine, California, ACM.
Davies, J. L. and T. R. Gurr (1998). "Preventive measures: Building risk assessment and crisis early warning systems", Rowman & Littlefield.
De Floriani, L., B. Falcidieno, et al. (1984). "A hierarchical structure for surface approximation." Computers & Graphics 8(2): 183-193.
De Groot, (2012). "Long-term coastal dune development in the Interactive Design Tool (HK 4.1)". Ecoshape Building with Nature Report
Deborah, J. A. (2006). "The quarks of object-oriented development." Commun. ACM 49(2): 123-128.
Deltares (2005). Introduction to Delft-FEWS.
Dilo, A. and S. Zlatanova "Spatiotemporal data modeling for disaster management in the Netherlands."
Donoghue, K. (2002). "Built for use: driving profitability through the user experience", McGraw-Hill.
Drabek, T. E. (2000). "The social factors that constrain human responses to flood warnings." Flood Hazards and Disasters vol. 1: 361–376.
Efraim, T. (1993). "Decision Support and Expert Systems: Management Support Systems", Prentice Hall PTR.
Engel, B. A., J.-Y. Choi, et al. (2003). "Web-based DSS for hydrologic impact evaluation of small watershed land use changes." Computers and Electronics in Agriculture 39(3): 241-249.

Erik Kjems, L. B. (2008). "Virtual Reality for training and collaboration in emergency management". Geospatial Information Technology for Emergency Response. J. L. Sisi Zlatanova.
Faber, B., Wallace, W., Cuthertson, J., (1995). "Advances in collaborative GIS for land resource negotiation". GIS '95 Symposium, Vancouver.
Faber, B., Watts, R., Hautaluoma, J., Wallace, W., Wallace, L., (1994). "A Groupware-enabled GIS". GIS '94 Symposium, Vancouver.
Fabian Herman, F. H. (2002). "User Requirement Analysis and Interface Conception for a Mobile, Location-Based Fair Guide". Human Computer Interaction with Mobile Devices, Berlin, Springer.
FEMA.gov. (2012). Federal Emergency Management Agency.
Fang, Y. I. N. and M. Feng (2009). "A WebGIS framework for vector geospatial data sharing based on open source projects."
Feng, M., S. Liu, et al. "Prototyping an online wetland ecosystem services model using open model sharing standards." Environmental Modelling & Software 26(4): 458-468.
Finkl, C. W. (2002). "Long-term analysis of trends in shore protection based on papers appearing in the Journal of Coastal Research, 1984-2000." Journal of Coastal Research: 211-224.
Finlay, P. N. (1989). "Introducting decision support systems", Oxford, NCC Blackwell.
Frøkjær, E., M. Hertzum, et al. (2000). "Measuring usability: are effectiveness, efficiency, and satisfaction really correlated?" CHI '00: Proceedings of the SIGCHI conference on Human factors in computing systems, ACM Press.
Frank Molkenthin, K. P. H. (1998). "Working Process in a Virtual Institute". Conference of Hydroinformatics.
Garcia, J. Ext JS in Action, Manning Publications.
Germs, R., G. Van Maren, et al. (1999). "A multi-view VR interface for 3D GIS." Computers & Graphics 23(4): 497-506.
Gineke Snoeren, S. Z., Joep Crompvoets, Henk Scholten (2007). "Spatial Data Infrastructure of emergency management: the view of the users".
Gong J.H., L. H. Z., Jian T. (2002). "Research on Geo-Process Based Computation Visualisation - A Case Studt of Flood Simulation." Journal of Chang'an University Vol.22: 186-190.
Goodchild, M. F. (2007). "Citizens as sensors: web 2.0 and the volunteering of geographic information." GeoFocus nº 7: 8-10.
Goosen, H., R. Janssen, et al. (2007). "Decision support for participatory wetland decision-making." Ecological Engineering 30(2): 187-199.
Graffy, E. A. a. N. L. B. (2008). "Linking Environmental Risk Assessment and Communication: An Experiment in Co-Evolving Scientific and Social

Knowledge." International Journal of Global Environmental(8(1/2)): 132-146.

Gregersen, J. B., et al. (2007). "OpenMI: Open Modeling Interface". Journal of Hydroinformatics 9(3). 175-191.

Grothe, M., H. Landa, et al. (2005). "The Value of Gi4DM for Transport & Water Management". Geo-information for Disaster Management: 129-153.

Guttler, R., R. Denzer, et al. (2001). "User interfaces for environmental information systems - interactive maps or catalog structures? Or both?" Advances in Environmental Research 5(4): 345-350.

Haklay, M., A. Singleton, et al. (2008). "Web mapping 2.0: The neogeography of the GeoWeb." Geography Compass 2(6): 2011-2039.

Handmer, J. "Flood warning reviews in North America and Europe: statements and silence."

Handmer, J. (2002). "Flood warning reviews in North America and Europe: statements and silence." Australian Journal of Emergency Management 17(3): 17–24.

Heng-Li Yang, J.-H. T. (2003). "A three-stage model of requirements elicitation for Web-based information systems." Industrial Management & Data Systems.

Herbert Alexander, S. (1977). "The New Science of Management Decision", Prentice Hall PTR.

Hermans, L. M. (2006). "Stakeholder-oriented valuation to support water resources management processes: Confronting concepts with local Practice", FAO.

Hoang Li. (2013). "The effect of riparian zones on nitrate removal by denitrification at the river basin scale". Phd thesis.

Hu, P. J.-H., P.-C. Ma, et al. (1999). "Evaluation of user interface designs for information retrieval systems: a computer-based experiment." Decision Support Systems 27(1-2): 125-143.

Huang, B. (2003). "Web-based dynamic and interactive environmental visualisation." Computers, Environment and Urban Systems 27(6): 623-636.

Hugues, H. (1998). "Smooth view-dependent level-of-detail control and its application to terrain rendering". Proceedings of the conference on Visualisation '98. Research Triangle Park, North Carolina, United States, IEEE Computer Society Press.

Hui Ma, A. E. M. (2007). "Assessing the Ecological Environment Around the Yellow River Delta Area by RS and GIS", The 3rd Yellow River Forum, Dong Ying, China.

Huib de Vried, A. W. (2009). "Building with Nature: ecodynamic design in practice". 2nd German Environmental Sociology Summit Reshapong Nature: Old Limits and New Possibilities.

Jankowski, P. (2009). "Towards participatory geographic information systems for community-based environmental decision making." Journal of Environmental Management 90(6): 1966-1971.

Jankowski, P. and M. Stasik (1997). "Spatial understanding and decision support system: A prototype for public GIS." Transactions in GIS 2(1): 73-84.

Janssen, M. A., H. Goosen, et al. (2006). "A simple mediation and negotiation support tool for water management in the Netherlands." Landscape and Urban Planning 78(1-2): 71-84.

Jern, M. (2005). "Web-Based 3D Visual User Interface to a Flood Forecasting System". Geo-information for Disaster Management: 1021-1039.

H. Xu, et al.(2010). "An instant and interactive platform based on Google Earth Plug-in". Advanced Computer Control (ICACC), 2010 2nd International Conference on.

Jianfeng Zhao, David J. Coleman. (2007). "An Empirical Assessment of a Web-based PPGIS prototype". URISA, Vol. 2007

Jin-Yong Choi, B. A. E. a. R. L. F. (2005). "Web-based GIS and spatial decision support system for watershed management." Journal of Hydroinformatics.

Jo, W., D. Jason, et al. (2007). "Interactive Visual Exploration of a Large Spatio-temporal Dataset: Reflections on a Geovisualisation Mashup", IEEE Educational Activities Department. 13: 1176-1183.

Jude, S. (2008). "Investigating the potential role of visualisation techniques in participatory coastal management." Coastal Management 36(4): 331-349.

Jude, S. R., A. P. Jones, et al. (2007). "The development of a visualisation methodology for integrated coastal management." Coastal Management 35(5): 525-544.

Karatzas, K., E. Dioudi, et al. (2003). "Identification of major components for integrated urban air quality management and information systems via user requirements prioritisation." Environmental Modelling & Software 18(2): 173-178.

Kardos, J., A. Moore, et al. "Visualising Uncertainty in Geographic Data using Hierarchical Spatial Data Structures."

Kemec, S., H. S. Duzgun, et al. (2009). "A conceptual framework for 3D visualisation to support urban disaster management. "

Kevin, S. (2008). "Thresholds. Flood Warning, Forecasting and Emergency Response".

Klimenko, S. V. B., Dmitry A. Danilicheva, Polina P. Fomin, Sergey A. Borisov, Tengiz N. Islamov, Rustam T. Kirillov, Igor A Lukashevich, Igor E. Baturin, Yury M. Romanov, Alexey A. Tsyganov, Sergey A. (2008). "Using Virtual Environment Systems During the Emergency Prevention, Preparedness, Response and Recovery Phases". Resilience of Cities to Terrorist and other Threats: 475-490.

Kmat, V. R., J.C.Martinez. (2005). "Large-Scale Dynamic Terrain In Three-Dimensional Construction " Journal of Computing in Civil Engeering Vol 19: 160-177.

Koller, D., P. Lindstrom, et al. (1995). "Virtual GIS: A real-time 3D geographic information system". Proceedings of the 6th conference on Visualisation'95, IEEE Computer Society.

Koningsveld, M. v. (2003). "Matching Specialist Knowledge with End User Needs", University of Twente. Doctor.

Kropla, B. (2005). "Beginning MapServer: open source GIS development", Apress.

Lam, D., L. Leon, et al. (2004). "Multi-model integration in a decision support system: a technical user interface approach for watershed and lake management scenarios." Environmental Modelling & Software 19(3): 317-324.

Lam, D. C. L., Wong, I., Swayne, D. A. and Fong, P. (1994). "Data and knowledge visualisation in an environmental information system." Journal of Biological Systems 2: 481-497.

Lattuada, R. (2006). "Chapter3: Three-Dimensional Representations and Data Structures in GIS and AEC".

Leon Hermans, D. R., Lucy Emerton, Danièle Perrot-Maître, Sophie Nguyen-Khoa, Laurence Smith (2006). "Stakeholder-oriented valuation to support water resources management processes". Rome.

Lorensen, B. (2004). "On the Death of Visualisation", NIH/NSF Fall 2004 Workshop on Visualisation Research Challenges.

M.J.Bezuyen, M., M.J.van Duin and P.H.J.A. Leenders (1998). "Flood management in The Netherlands." Australian Journal of Emergency Management: 43-49.

M.J.North, T. R. H., N.T.Collier, J.R.Vos (2005). "The Repast Simphony Development Environment". Agent 2005 Conference on Generative Social Process, Models, and Mechanisms, Chicago.

MacEachren, A. M. (1991). "The role of maps in spatial knowledge acquisition." The Cartographic Journal 28: 10.

Mahesh, R., F. Guoliang, et al. (2007). "A web-based GIS Decision Support System for managing and planning USDA's Conservation Reserve Program (CRP)", Elsevier Science Publishers B. V. 22: 1270-1280.

Mari, R., L. Bottai, et al. (2011). "A GIS-based interactive web decision support system for planning wind farms in Tuscany (Italy)." Renewable Energy 36(2): 754-763.

Mark, D., W. Murray, et al. (1997). "ROAMing terrain: real-time optimally adapting meshes". Proceedings of the 8th conference on Visualisation '97. Phoenix, Arizona, United States, IEEE Computer Society Press.

Mark, D. M., J. P. Lauzon, et al. (1989). "A review of quadtree-based strategies for interfacing coverage data with digital elevation models in grid form." International journal of geographical information systems 3(1): 3-14.

Matthies, M., C. Giupponi, et al. (2007). "Environmental decision support systems: Current issues, methods and tools." Environmental Modelling & Software 22(2): 123-127.

McIntosh, B. S., J. C. Ascough Ii, et al. (2011). "Environmental decision support systems (EDSS) development - Challenges and best practices." Environmental Modelling & Software 26(12): 1389-1402.

Meissner, A., T. Luckenbach, et al. (2002). "Design challenges for an integrated disaster management communication and information system". The First IEEE Workshop on Disaster Recovery Networks (DIREN 2002).

Moggridge, Bill (2007). "Designing Interactions". MIT Press. ISBN 0-262-13474-8.

Morton, P. G. W. K. a. M. S. S. (1978). "Decision Support Systems: An Organizational Perspective".

Mynett, A. E., I.A. Sadarjoen, A.J.S. Hin (1995). "Turbulent flow visualisation in computational and experimental hydraulics", Visualisation '95, Los Alamitos.

Neches, R., R. E. Fikes, et al. (1991). "Enabling technology for knowledge sharing." AI magazine 12(3): 36.

Nonaka, I. (1994). "A Dynamic Theory of Organizational Knowledge Creation." Organization Science 5(1): 14-37.

Nielsen, J. (1994). Heuristic evaluation. In Nielsen, J., and Mack, R.L. (Eds.), "Usability Inspection Methods", John Wiley & Sons, New York

Norliza Katuk, K.-K. R. M., Norita Norwawi, Safaai Deris (2009). "Web-based support system for flood response operation in Malaysia." Disaster Prevention and Management 18(3): 327-337.

Norman Donald (1988). "The Design of Everyday Things". New York: Basic Books. ISBN: 978-0-465-06710-7

North, M. J. and C. M. Macal (2009). "Foundations of and Recent Advances in Artificial Life Modeling with Repast 3 and Repast Simphony". Artificial Life Models in Software: 37-60.

Open, G. L. A. R. B., S. Dave, et al. (2007). "OpenGL Programming Guide: The Official Guide to Learning OpenGL", Version 2.1, Addison-Wesley Professional.

Orland, B., K. Budthimedhee, et al. (2001). "Considering virtual worlds as representations of landscape realities and as tools for landscape planning." Landscape and Urban Planning 54(1-4): 139-148.

Oztekin, A., A. Nikov, et al. (2009). "UWIS: An assessment methodology for usability of web-based information systems." Journal of Systems and Software 82(12): 2038-2050.

Pajarola, R. (1998). "Large scale terrain visualisation using the restricted quadtree triangulation". Visualisation'98. Proceedings, IEEE.

Parker, D. J. (2003). "Designing Flood Forecasting, Warning and Response Systems from a Societal Perspective". International Conference on Alpine Meteorology and Meso-Alpine Programme. Brig, Switzerland.

Paul Graham. (2005). "Web 2.0"

Pereira, A. G., Quintana, S.C., (2002). "From technocratic to participatory decision support systems responding to the new governance initiatives." Journal of Geographic Information and Decision Analysis Vol 6: 95-107.

Peter, L., K. David, et al. (1996). "Real-time, continuous level of detail rendering of height fields". Proceedings of the 23rd annual conference on Computer graphics and interactive techniques, ACM.

Pietro, T. "Economic Simulations in Swarm: Agent-Based Modelling and Object Oriented Programming - By Benedikt Stefansson and Francesco Luna: A Review and Some Comments about Agent Based Modeling." The Electronic Journal of Evolutionary Modeling and Economic Dynamics.

Pilon, P. J. E. (2006). "Key Elements of Flood Disaster Management Guidelines for reducing flood losses", United Nation.

R K Price, A. E. M. (2002). "Knowledge management in a virtual research organisation: consequences for education and research in hydroinformatics". Hydroinformatics 2002, Cardiff, UK, IWA Publishing and the authors.

R. Soncini-Sessa, A. N., C. Gandolfi, A. Kraszewski (1990). "Computeraided water reservoir management: a prototype two level DSS", Invited paper: NATO ARW on ComputerAided Support Systems in Water Resources Research and Management. Ericeira, Portugal.

R.W.Greene (2002). "A GIS Handbook", ESRI Press.

Ralf, D. (2005). "Generic integration of environmental decision support systems - state-of-the-art." Environmental Modelling & Software 20(10): 1217-1223.

Renate Steinmann , A. K., Thomas Blaschke (2004). "Analysis of Online Public Participatory GIS Applications with Respect to the differences between the US and Europe". 24TH URBAN DATA MANAGEMENT SYMPOSIUM.

Research, C. I. o. W. R. a. H. (2006). "Digital River Hand Book".

Richman, A. (2008). "Virtual Environmental Planning system" (VEPs).

Rinner, C., C. Keßler, et al. (2008). "The use of Web 2.0 concepts to support deliberation in spatial decision-making." Computers, Environment and Urban Systems 32(5): 386-395.

Rizzoli, A. E. and W. J. Young (1997). "Delivering environmental decision support systems: software tools and techniques." Environmental Modelling & Software 12(2-3): 237-249.

Ronald, J. N. (1996). "Object-oriented systems analysis and design", Prentice-Hall, Inc.

Rottger, S., W. Heidrich, et al. (1998). "Real-Time Generation of Continuous Levels of Detail for Height Fields". Proc. 6th Int. Conf. in Central Europe on Computer Graphics and Visualisation.

Samet, H., A. Rosenfeld, et al. (1984). "A geographic information system using quadtrees." Pattern Recognition 17(6): 647-656.

Scholten, H. J., L. Andrea, et al. (1998). "Towards a spatial information infrastructure for flood management in The Netherlands." Journal of Coastal Conservation 4(2): 151-160.

Schulte, C. (2008). "Concept of a CityGML Application Domain Extension for a web based 3D flood information service". Photogrammetry and Geoinformatics Stuttgart University of Applied Sciences. Master.

Sene, K. (2008). "Flood warning, forecasting and emergency response". Berlin Springer 2008.

Shaffer, C. A., R. Juvvadi, et al. (1993). "Generalized comparison of quadtree and bintree storage requirements." Image and Vision Computing 11(7): 402-412.

Shaffer, C. A., R. Juvvadi and L.S. Heath. (1993). "Generalized comparison of quadtree and bintree storage requirements." Image and Vision Computing Vol 11(7): 402-412.

Shi, P., J. a. Wang, et al. (2000). "Understanding of natural disaster database design and compilation of digital atlas of natural disasters in China." Geographic Information Sciences 6(2): 153-158.

Shi, S., X. Ye, et al. (2007). "Real-time simulation of large-scale dynamic river water." Simulation Modelling Practice and Theory 15(6): 635-646.

Shim, J. P., W. Merrill, et al. (2002). "Past, present, and future of decision support technology." Decis. Support Syst. 33(2): 111-126.

Simmonds, P., J. Stroyan, et al. (2004). "Survey of knowledge transfer between NERC-funded researchers and the users of their outputs".

Smith, T. M. and V. Lakshmanan (2006) "Utilizing Google Earth as a GIS platform for weather applications". 22nd International Conference on Interactive Information Processing Systems for Meteorology, Oceanography, and Hydrology.

Snoeren, G. (2007). "User requirements for a Spatial Infrastructure for Emergency Management", Utrecht Univ., TU Delft, Wageningen Univ., ITC. MSC: 179 p.

Sprague, R. H. and E. D. Carlson (1982). "Building effective decision support systems". Englewood Cliffs, N.J., Prentice-Hall.

Taal, M. M., J., Cleveringa, J., Dunsbergen, D. (2006). "15 years of coastal management in the Netherlands, Policy; Implementation and Knowledge Framework." Rijkswaterstaat, National Institute for Coastal and Marine Management/RIKZ.

Thomas Kolbe, S. B. (2006). "CityGML: An Open Standard for 3D City Models". Directions Magazine.

Trygve Reenskaug, James Coplien. (2009). "The DCI Architecture: A New Vision of Object-Oriented Programming".

Tsichritzis, D. a. F. L. (1982). "Data Models", Prentice Hall.

Van Koningsveld, M. D. B., G.J. Baart, F. Damsma, T. Den Heijer, C. Van Geer, P. De Sonnevile, B. (2010). "OpenEarth - Inter-Company Management of: Data, Models, Tools & Knowledge". WODCON XIX Conference: Dredging Makes the World a Better Place, Beijing.

Verbree, E., G. V. Maren, et al. (1999). "Interaction in virtual world views-linking 3D GIS with VR." International Journal of Geographical Information Science 13(4): 385-396.

Wasfy, T. M. and A. K. Noor (2002). "Rule-based natural-language interface for virtual environments." Advances in Engineering Software 33(3): 155-168.

Wheaton, J. M., C. Garrard, et al. "A simple, interactive GIS tool for transforming assumed total station surveys to real world coordinates the CHaMP transformation tool." Computers & Geosciences 42: 28-36.

Wijk, J. J. v. (2005). "The Value of Visualisation". 16th IEEE Visualisation Conference VIS 2005. Minneapolis, MN, USA, IEEE Computer Society: 11.

Wu Dong, A. E. M. (2007). "Evaluation of the flow regime in the Lower Yellow River Delta", China. The 3rd Yellow River Forum, Dong Ying, China.

Xu, W. and S. Zlatanova (2007). "Ontologies for disaster management response". Geomatics Solutions for Disaster Management, Springer: 185-200.

Xuan Zhu, A. E. M. (2009). "3D Visualisation and Virtual Reality Tools for accelerating decision making processes". In Proceedings of the 8th International Conference on Hydroinformatics, Conception, Chile.

Zafiri, M. M. H. a. A. (2008). "Usability Engineering for GIS: Learning from a Screenshot." The Cartographic Journal Vol. 45 No. 2(Use and Users Special): pp. 87–97.

Zhang, C., H. Tang, et al. (2009). "Web-Based Virtual Environment for Decision Support System in Water Based System". Advances in Water Resources and Hydraulic Engineering, Springer Berlin Heidelberg: 545-550.

Zhang, J. (1997). "The nature of external representations in problem solving." Cognitive Science: A Multidisciplinary Journal 21(2): 179 - 217.

Zhang, L., C. Yang, et al. (2007). "Visualisation of large spatial data in networking environments." Computers & Geosciences 33(9): 1130-1139.

Zheng Xu, S. Y., Ann van Griensven and Linh Hoang (2011). "Automated building of Bayesian Belief Networks to replicate the spatial and ecotoxicological relationships in river basins". the SWAT-SEA conference in Ho Chi Minh city. Vietnam.

Zhou, X. (2006). "Research on Semantics-based Software Component Retrieval Method and its Application in the Domain of Water Resources". Institute of Hydraulic and Hydro-Power Engineering. Nanjing, Hohai University Doctor of Engineering: 129.

Zhu, B. and H. Chen (2005). "Using 3D interfaces to facilitate the spatial knowledge retrieval: a geo-referenced knowledge repository system." Decision Support Systems 40(2): 167-182.

Zhu, B., M. Ramsey, et al. (1999). "Support concept-based multimedia information retrieval: a knowledge management approach". Proceedings of the 20th international conference on Information Systems, Association for Information Systems.

Zhu Q., L. F. C., Zhang Y.T., (2007). "Unified representation of 3D city models." Journal of Chang'an University Vol 27: 54-58.

Zlatanova, M. B. a. S. (2004). "3D Geo-DBMS".

Zlatanova, S. "SII for emergency response: the 3D challenges."

Zlatanova, S. (2000). "3D GIS for urban development", International Institute for Aerospace Survey and Earth Sciences.

Zlatanova, S. (2008). "SII for emergency response: the 3D Challenges". The international Archives of Photogrammery, Romote Sensing and Spatial Information Sciences Beijing.

Zlatanova, S. P., David. (2006). "Large-scale 3D Data Integration: Challenges and Opportunities ", CRCpress, Tailor & Francis Group, Boca Raton.

About the Author

Xuan Zhu was born in Nanjing, Jiangsu Province, China on 1st October 1982. She studied computer science at Nanjing University of Technology from 2001 to 2005 and graduated with a bachelor degree in multimedia science. She then joined a collaborative masters' programme in Hydroinformatics between Hohai University and UNESCO-IHE. She followed the MSc programme in the Netherlands from 2006 to 2008 and graduated with the Master of Science degree. She then continued to pursue her PhD degree at UNESCO-IHE under the supervision of Professor Arthur Mynett, sponsored by Deltares. She is presently working as a software engineer at Allseas Engineering B.V. in Delft, working for the oil and gas industry.

She married with Hui Chen and got their daughter Yisi Chen in 2010.

Publications

Zhu, X ; Mynett, A. E. (2007). 3D Graphics and Virtual Reality Applications in Decision Support System. In *Proceedings of the 3rd International Yellow River Forum on Sustainable Water Resource Management and Delta Ecosystem Maintenance, ISBN: 978-780-734-2960,* (pp. 311-322)

Zhu, X ; Mynett, A. E. (2009). Web-Based Virtual Environment for Decision Support System in Water Based System. *Advances in Water Resources and Hydraulic Engineering,Vol.II, ISBN: 978-3-540-89465-0,* (pp. 545-550)

Zhu, X ; Mynett, A. E. (2009). 3D Visualisation and virtual reality tools for accelerating decision making processes. In *Proceedings of the 8th International conference on Hydroinformatics, Conception, Chile, ISBN: 978-1-617-82059-5* (pp. 1181-1190)

Zhu, X ; Mynett, A. E. (2009). An Interactive Virtual Environment for Assessing the Ecohydraulics in the Lower Part of Yellow River Delta. In

Proceedings 33rd IAHR Congress: Water Engineering for a sustainable Environment, ISBN: 978-1-61738-231-4 (pp. 3800-3807).

Zhu, X ; Mynett, A. E. (2009). Appropiate user interface for decision support systems in water resource management. In *Proceedings 33rd IAHR Congress: Water Engineering for a sustainable Environment, ISBN: 978-1-61738-231-4* (pp. 5302-5308).

Zhu, X ; Li, H. ; Mynett, A. E. (2010). Using concept of object-oriented programming for Aquatic model development. In Tao Jianhua, Chen Qiuwen & Liong Shie-Yui (Eds.), *Proceedings of the 9th International conference on Hydroinformatics, Tianjin, China* (pp. 2550-2557). Beijing: Chemical Industry Press.

Zhu, X ; Mynett, A. E. (2010). Using UML models for flood disaster management system design focus on 3D spatial information dissemination. In Tao Jianhua, Chen Qiuwen & Liong Shie-Yui (Eds.), *Proceedings of the 9th International conference on Hydroinformatics, Tianjin, China* (pp. 1825-1832). Beijing: Chemical Industry Press.

Zhu, X ; Mynett, A. E. ; Xuan,Y ; Van Griensven. A (2012). Web-based FOSS Environment Decision Support Systems for Odense River Basin in Demark. *Proceedings of the 6th biannual meeting of International Environmental Modelling and Software Society, Leipzig, Germany ISBN: 978-88-9035-742-8* (pp. 17-24).

Zhu, X ; Mynett, A. E. (2012). One-Page-One-System Decision Support Tool for Water Resource Management. *International Journal of Environment Modeling and Software* (under review).

Zhu, X ; Mynett, A. E. (2013). Arranging 3D information in Flood Disaster Management Systems. *Journal of Hydroinformatics* (in preparation).

Zhu, X ; Mynett, A. E. (2013). Web-based Public Collaborative Platform for Coastal Planning. *International Journal of Environment Modeling and Software* (in preparation).